NUCLEAR ENERGY

PERCEPTION OR REALITY?

Emrah Akyüz

NUCLEAR ENERGY

PERCEPTION OR REALITY?

Emrah Akyüz

Academica Press
Washington~London

Library of Congress Cataloging-in-Publication Data is available.

Names: Akyüz, Emrah (author)
Title: Nuclear energy : perception or reality? | Akyüz, Emrah
Description: Washington : Academica Press, 2023. | Includes bibliographical references and index.
Identifiers: LCCN 2022944786 | ISBN 9781680536928 (hardcover) | 9781680536935 (ebook)

I dedicate this book to

my mother Kibriye Akyüz and my father Ismail Akyüz.

CONTENTS

LIST OF TABLES

LIST OF FIGURES

CHAPTER I:
INTRODUCTION

1. RESEARCH BACKGROUND

Energy is one of the most important needs of human beings. Energy is regularly consumed at many stages of the lives of modern-day people. Energy is considered as one of the most important factors that play a decisive role in the improvement of living standards, the development of the economy, ensuring security, the advancement of technology, the realization of comfortable and safe transportation, and ensuring communication. Without energy, it is not possible for modern societies to maintain and develop their lifestyles. Therefore, energy is one of the most valuable and important tools of the 21st century world. However, the use of energy, which has strategic importance in today's world, is not a new phenomenon. Energy has been used to facilitate human life since the primitive period.

The history of energy use dates back to ancient times. The energy that people first discovered is renewable energy. The use of wind and sun as energy sources has thousands of years of history. With the use of the power of the wind, goods were transported for commercial purposes by boats and ships. In primitive times, ground mills were used to pump water and grind wheat, but wind power was also used. Wind energy appears to be the most widely used energy source of the primitive period. On the other hand, solar energy was directly utilized. The sun's rays are used for heating, especially in cold seasons. However, it is not possible to talk about technological developments that can convert renewable energy sources into electricity in these periods.

After renewable energy sources, another energy source that has been used since ancient times is fossil fuels. The use of fossil fuels, which is the most important energy source that meets the world's energy needs after the Industrial Revolution, has thousands of years of history. Mankind has been using oil and coal for thousands of years for different purposes. Primitive people burned oil and coal for heating, cooking, and security purposes. The environmental problems that emerged due to the excessive use of coal in Europe in the First Age and the Middle Ages reached deadly dimensions. For example, in order to prevent deaths caused by air pollution caused by excessive coal use in England, King Edward I, who ruled England from 1271 to 1307, made legal regulations prohibiting the use of coal.[1]

The turning point in the history of energy has been the Industrial Revolution. With the economic and social transformation experienced after the Industrial Revolution, the use of fossil fuels has expanded further. With the use of steam-powered machines in production and transportation, especially coal has started to gain more importance. In addition to coal, oil and natural gas consumption have also increased rapidly with technological developments. This momentum, which started with the Industrial Revolution, continued increasingly until the 1950s. Environmental problems caused by excessive consumption of fossil fuels and their effects on public health have triggered an increase in social reaction and a tendency to alternative energy sources, especially in European countries and the USA, since the 1950s. One of these alternative energy sources is nuclear energy.

Since the second half of the 20th century, nuclear energy has been at the center of many countries' energy policies because fossil fuels are a non-renewable energy source, difficulty accessing reserves due to their availability in certain geographical areas and most importantly, the devastating damage to the environment. During the Cold War, nuclear energy technology experienced rapid progress as the competition between the Western Bloc and the Soviet Union took place in the field of nuclear energy. Advances in nuclear energy technology have made it easier for countries to access this energy source.

The most important factors determining the nuclear energy perspective of countries have been energy crises and nuclear power plant

---

[1] David Urbinato, "London's Historic" Pea-soupers." (Smog in London, England)," *EPA Journal*, 20/1-2 (1994): 44-45.

accidents. In the 1970s, the tendency to alternative energy sources to fossil energy sources such as nuclear energy increased more due to the oil crises[2] that deeply affected the industrialized countries. Countries have started to invest more in nuclear energy in order to meet their energy needs and to ensure energy security. However, the trend toward nuclear energy has come to a standstill or decreasing with the accidents in nuclear power plants. The most important nuclear power plant accident that affects the energy policy of countries against nuclear energy is the Chernobyl Nuclear Power Plant accident that occurred in Ukraine in 1986. The environmental pollution that emerged with the Chernobyl disaster and its chaotic effects on living things have led to more questioning about whether nuclear energy is dangerous for the environment and public health.

With the Chernobyl disaster, while the use of nuclear energy was questioned more, on the other hand, security measures in the field of nuclear energy were tried to be increased more. In order to prevent nuclear power plant accidents like Chernobyl, nuclear technology has made remarkable progress in terms of safety. The view that nuclear power plants, which are protected with high-security measures, do not pose a risk in terms of the environment and public health, gained more prevalence towards the beginning of the 21st century with the reduction of the risk of accidents. However, the public support that has developed in favor of nuclear energy has tended to decrease again with the Fukushima Nuclear Power Plant accident that took place in Japan in 2011. In the 25 years between the Chernobyl disaster in 1986 and the Fukushima disaster in 2011, despite the important developments in nuclear energy technology, the failure to prevent nuclear power plant accidents has led to an increase in nuclear energy opposition all over the world.

Judging nuclear energy only by the environmental pollution caused by nuclear power plant accidents and its effects on living life gives rise to a one-sided perspective. Because when we look at the other side of the coin, it is seen that nuclear energy has significant advantages over other energy sources. One of the most important arguments of nuclear energy

---

[2] Olof Johansson & Leo Schipper, "Measuring the Long-Run Fuel Demand of Cars: Separate Estimations of Vehicle Stock, Mean Fuel İntensity, and Mean Annual Driving Distance," *Journal of Transport Economics and Policy*, 31/3 (1997): 277-292.

advocates is the view that nuclear energy can be a successful tool in the fight against climate change, which is considered the most important environmental problem of our time. The view that nuclear energy, which does not directly cause the intake of greenhouse gases during energy production, can be an effective instrument in the fight against climate change triggered by greenhouse gases caused by the use of fossil fuels is accepted by many people.

Nuclear energy is the most discussed energy source of the 21st century. Although nuclear energy is an energy source that has both advantages and disadvantages in terms of environment, public health and economy, nuclear energy opponents and advocates look at only one side of the coin and try to channel the society in this direction. The inadequacy of objective scientific studies on nuclear energy and the excess of subjective studies increase the manipulation of nuclear energy even more. The one-sided view of nuclear energy opponents and advocates of nuclear energy in visual and written media triggers the problem to deepen. All these processes cause information pollution about nuclear energy at the social level and affect the attitudes of individuals towards nuclear energy. For this reason, there is a need for scientific studies that take nuclear energy objectively and examine both sides of the coin, namely its advantages and disadvantages, at the point of enlightening society.

The main purpose of this study is to examine the advantages and disadvantages of nuclear energy in terms of environment, economy and public health, opportunities and risks, advantages and disadvantages, etc. In this scientific work, two basic research questions are tried to be answered. First, what are the advantages and disadvantages of nuclear energy in terms of environment[3], public health and economy? Second, what are the factors affecting individuals' attitudes towards nuclear energy?

---

[3] In this study, the environment is considered as a whole. The environment has both physical and biological elements. The physical elements of the environment are inanimate entities. Sea, ocean, forest, earth, air etc. form the physical elements of the environment. The biological elements of the environment include all living things.

2. THE SCOPE OF THE RESEARCH

There are many scientific studies on nuclear energy at the international level. Nuclear energy is investigated from different perspectives in both positive sciences and social sciences. Nuclear energy is most commonly studied by disciplines such as economics, environmental science, sociology, law, and political science. However, most of these studies fall short on two points. First, when dealing with nuclear energy, only one side of the coin is looked at or tried to be brought to the fore. Proponents of nuclear energy generally focus only on the advantages of nuclear energy, while opponents of nuclear energy focus on the disadvantages of nuclear energy.

Nuclear energy is an energy source with both advantages and disadvantages. Nuclear energy is neither perfect, as its proponents claim, nor is it a source of energy that we must look at with fear, as its opponents have stated. As with any energy source, nuclear energy has opportunities and risks for society, the economy and the environment. Scientific studies need to objectively show both sides of the coin, that is, examine both the pros and cons of nuclear energy. This study is neither nuclear energy advocacy nor opposition to nuclear energy. One of the original aspects of this work is that it allows individuals to learn nuclear energy in all its dimensions, that is, to present both sides of the coin to readers in an objective framework.

One of the other important shortcomings of existing studies is that they are not comprehensive enough. In order to better understand nuclear energy, first of all, it is necessary to examine the energy issue in detail. Nuclear energy is an energy source that has emerged to meet the increasing energy demand. For this reason, focusing solely on nuclear energy without discussing the importance of energy in human life and the economy in modern societies, the scope and environmental effects of existing energy sources, the pros and cons of other energy sources, constitutes an obstacle to a healthy analysis of the subject. This study first examines energy, the importance of energy, the use of energy in the historical process, energy resources and the world's energy view from different dimensions, and then discusses the opportunities and risks of nuclear energy with an inclusive approach. This book allows readers to make sense of nuclear energy within the energy context.

The most original aspect of this study is the analysis of nuclear energy in the light of previously unpublished data. There have been two major nuclear power plant accidents in the world. The first of these is the Chernobyl Nuclear Power Plant accident that occurred in Ukraine in 1986. The second is the Fukushima Nuclear Power Plant accident that took place in Japan in 2011. The two most serious nuclear power plant accidents in the history of nuclear energy are very important for our understanding of nuclear energy. However, scientists who do scientific studies on nuclear energy do not have a comprehensive study of these two nuclear disasters. The author of this book, Dr. Emrah Akyuz, within the scope of his doctorate at the University of Leeds, England, has conducted long-term scientific research in the region where both the Chernobyl and Fukushima nuclear power plant accidents occurred.

As the first and only Turkish scientist to have conducted scientific studies in the regions of Fukushima and Chernobyl, where the largest nuclear power plant accidents occurred in our history, this book discusses nuclear energy and uses original data obtained from field studies conducted in 2017. Visual materials about Chernobyl and Fukushima accidents that have not been shared or used elsewhere before are presented to the readers for the first time in this work. One of the most important disadvantages of nuclear energy is the possible accident risk and its environmental and social effects. The materials and information presented in this study about the Fukushima and Chernobyl accidents allow us to better understand the extent of the accidents to be experienced in nuclear power plants.

3. TARGET AUDIENCE

Nuclear energy is widely discussed by four main groups which constitute the main target audience of this book. The first of these are scientists and researchers. Nuclear energy is not adequately studied in the field of social sciences. The social, environmental and economic dimensions of nuclear energy are important issues that need to be addressed. There are either not enough studies in this area or studies generally examine events and cases from a single window. In other words, scientific studies in the field of social sciences related to nuclear energy generally focus on one side of the coin, that is, only or heavily, on its advantages or disadvantages. For this reason, there is a need for objective works that deal with nuclear energy from different perspectives for social scientists and researchers working in the field of nuclear energy. This book has been created as a work that scientists and researchers working in the field of nuclear energy can benefit from in their scientific studies.

The second main target audience of this book is politicians and bureaucrats who shape energy policies. Which energy sources in a country and how they will be used are determined by the policies created by the decision-makers. Building or not building nuclear power plants in a country is determined only by the initiative of the politicians who rule the country. It is a situation seen in modern democracies that this decision is taken by the people by taking it to a referendum. But both situations depend on the will of the politicians. For this reason, it is very important for politicians and bureaucrats working in this field to have objective and internationally accepted information on topics such as energy, energy resources, nuclear energy, advantages and disadvantages of nuclear energy, in order to take rational and healthy decisions about energy policies. This book is a scientific study that allows politicians and bureaucrats to have a comprehensive knowledge of energy and nuclear energy.

The third target audience of this scientific study is non-governmental organizations. Nuclear energy is one of the main areas of interest of many non-governmental organizations in the world. Especially environmentalist non-governmental organizations show great interest in this issue due to the environmental effects of nuclear energy. It is possible to see the demonstrations and works of non-governmental organizations on nuclear energy in many parts of the world. However, it is questionable to what extent non-governmental organizations act objectively on this issue. Discourses that focus on only one side of nuclear energy and the activities that result from it are the subject of criticism. The main reason for the existence of non-governmental organizations is not to oppose or advocate nuclear energy, but to encourage politicians to make decisions for the benefit of society and nature. It will be beneficial for society to raise awareness of non-governmental organizations with scientific and consistent information on nuclear energy. This study includes studies that can contribute to the scientific knowledge of non-governmental organizations working in the field of nuclear energy about energy and energy-environment relationship.

The issue of nuclear energy is also discussed in society, away from the scientific framework. Individuals tend to view nuclear energy from a political perspective. Nuclear energy is not discussed over the opportunities and risks posed by this energy source in terms of society, economy, and environment. The number of individuals who evaluate nuclear energy through the discourses of the political parties they support is quite high. Especially those living around nuclear power plants determine their attitudes towards nuclear energy with the discourses they hear from the media and the politicians around them. This situation causes individuals to be manipulated or misled about nuclear energy. For this reason, another of the main target groups of this scientific study is individuals who have been manipulated with unscientific information about nuclear energy or consciously directed for certain political goals.

Another platform where nuclear energy is talked about the most is the media sector. Nuclear energy is discussed and examined from different perspectives in the written and visual media. However, it is not possible to say that those working in the media have comprehensive and impartial knowledge about nuclear energy. Since the media is one of the most important influence mechanisms that guide and provide information to the society with their news and studies, it seems important for media workers to have objective information on nuclear energy in terms of informing the society correctly. Therefore, another target audience of this book is those who work in written and visual media.

4. SUMMARY OF CHAPTERS

This study consists of five main parts. In the second chapter titled "*Energy*," the concept of energy is examined with its different dimensions. Although energy is a concept that we use frequently in our daily lives, the dimensions of this concept are quite wide. For this reason, in the second chapter, firstly, what energy means, the scope of energy, and the importance of energy in human life are examined. Then, the types of energy sources are examined. Energy is not obtained from a single source. The sources of energy are divided into three groups as fossil fuels, nuclear energy, and renewable energy sources. In this section, the energy sources and their scope are discussed in detail. In the third part, energy history is evaluated. The energy resources used in human history and how they evolved are evaluated. Finally, the energy outlook of the world is examined. The world's energy needs are increasing day by day. It is discussed which energy sources are used to meet the increasing energy need.

In the third chapter titled "*Nuclear Energy*," the scope of nuclear energy is discussed. In this chapter, descriptive information about the scope of nuclear energy is given. The third chapter consists of three parts. In the first part, what nuclear energy is, the main raw material of nuclear energy, and how nuclear energy is produced are evaluated. Afterwards, the usage areas of nuclear energy are examined. In the second part, the history of nuclear energy is discussed. Nuclear energy is relatively new compared to other energy sources. This chapter examines how nuclear energy emerged and developed. In the last part, the use of nuclear energy in the world is analyzed. Nuclear energy is used by 32 countries. However, the number of countries using nuclear energy tends to increase. Turkey and Bangladesh are building their first nuclear power plants. In this section, how nuclear energy has evolved in the energy policy of countries is discussed.

In the fourth chapter titled "*Advantages of Nuclear Energy*," the advantages of nuclear energy in terms of society, environment, and economy are analyzed. Nuclear energy has many advantages. It offers significant opportunities against both fossil fuels and renewable energy sources. The advantages of nuclear energy discussed in this chapter are as follows; (1) long operating lifetime of nuclear power plants, (2) low level of greenhouse gases caused by nuclear energy, (3) nuclear power plants not triggering the release of pollutants into the air, (4) high fuel efficiency of nuclear energy, (5) nuclear energy provides energy security, (6) nuclear energy contributes to the increase in employment by creating job opportunities at the regional level, (7) requires a low level of land use, and (8) contributes to reducing foreign dependency on energy.

In the fifth chapter titled "*Disadvantages of Nuclear Energy*," what kind of risks nuclear energy poses for the environment, society and economy are analyzed. In this chapter, the disadvantages of nuclear energy and nuclear power plants are examined by comparing them with other energy sources. The disadvantages of nuclear energy discussed in this section are as follows; (1) the environmental cost of nuclear energy is high, (2) nuclear energy is not sustainable, (3) there is a risk of producing nuclear weapons from nuclear energy, (4) nuclear wastes generated after energy production in nuclear power plants, (5) nuclear energy has the higher the installation cost of power plants compared to other energy sources, (6) the risk of being exposed to terrorist attacks.

CHAPTER II:
ENERGY

1. INTRODUCTION

Energy is the workforce needed to change the state/position of anything that exists physically in nature. Energy has been used and continues to be used by mankind since the very earliest periods of history. One of the most basic needs of modern-day people in particular is energy. Energy is needed in many stages of our daily lives. Energy is consumed in all physical spaces where human beings exist. Energy is widely used in residential areas, social areas, and industrial facilities.

The energy that people need is met by different sources in nature. Energy is obtained by converting natural resources. Fossil fuels are the most widely used sources of energy. Most of the world's energy needs are met by fossil fuels such as petroleum and coal, which contain high carbon and are found in limited quantities in nature. The second main energy source is renewable energy. Energy is obtained from sources such as wind, sun, and waves, which are inexhaustible. The third most common energy source is nuclear energy. It is based on the energy released by the fission of uranium.

The main purpose of this chapter is to present general information about energy. This section is divided into five parts. In the first part, what energy is, the importance of energy, and its usage areas are examined. In the second part, types of energy sources are discussed. What energy sources are, their basic features, and their effects on the environment are argued. In the third part, the subject of energy history is discussed. How energy resources are used in the historical process is examined. In the last part, the energy outlook of the world is analyzed.

2. ENERGY

Energy, a concept we often use in our daily lives, is a Greek term as a word origin.[4] The concept of energy comes from the combination of the Greek words "en" meaning "within" and "ergy" meaning "work." Energy, according to its dictionary meaning, can be defined as the power or capacity to do a job.[5,6] According to the Cambridge English Dictionary, energy is conceptualized as the power obtained from a source to do a job.[7] According to the Turkish Language Association, energy is expressed as follows; "*The power that exists in the matter and emerges in the form of heat and light is energy.*"[8]

In order to understand more concretely what energy means, it may be more decisive to categorize energy according to its sources. There are different forms of energy according to its sources.[9] The first and most widely used of these is energy based on muscle power. It is an energy that human beings have used since their existence. Energy based on muscle power is used in all phases of our daily life. For example, moving an item from one place to another by using only our own limbs or moving our own body from one place to another is done with energy based on muscle power.

The second basic form of energy emerges with the physical elements of nature itself. The main source of energy is natural resources. The source of energy is constantly and periodically renewing itself. Renewable energy sources such as solar energy, wind energy, wave energy, and geothermal energy that exist in nature can be evaluated in this energy form. To explain with a more concrete example, the transport of a physical object from one place to another by the wind or the waves in the

---

[4] Cara New Daggett, The Birth of Energy: Fossil Fuels, Thermodynamics and the Politics of Work (Durham: Duke University Press, 2019),17.

[5] Matthew T. Huber, "Energizing Historical Materialism: Fossil Fuels, Space and the Capitalist Mode of Production," *Geoforum* 40/1 (2009): 105-115,105.

[6] Matthew T. Huber and James McCarthy, "Beyond the Subterranean Energy Regime? Fuel, Land Use and the Production of Space," *Transactions of the Institute of British Geographers* 42/4 (2017): 655-668, 656.

[7] "Meaning of Energy in English," last updated date: 15 January, 2022, https://dictionary.cambridge.org/dictionary/english/energy

[8] "Güncel Türkçe Sözlük," last updated date: 15 January, 2022, https://sozluk.gov.tr/

[9] Matthew T. Huber, "Energizing Historical Materialism," 106.

water resources occurs by using or transforming the energy that the nature reveals with its own physical elements.[10]

The energy that emerges with the physical elements of nature can be transformed with the technological tools invented by humans. Especially after the Industrial Revolution, the technological tools that people have obtained by using natural resources have facilitated the conversion of energy resources in nature. The most widely used energy in the modern period is the energy obtained or transformed by the use of machines invented by humans.[11] Machinery used in industrial facilities, all technological products used in residences or vehicles are included in this scope.

The third basic form of energy is the energy obtained by the use of the own power of other living species other than humans in nature. As in humans, animals also have energy based on muscle power. Human beings have been using this energy for their own interests or needs since ancient times. One of them is its use as a means of transportation.[12, 13] For example, it was very common to make long journeys using animals such as horses, dogs, and donkeys, before the invention of vehicles working with fossil fuels or renewable energy sources. Using the physical strength of animals to do a job is still common in rural areas.

Energy, which is used in different forms, is one of the most important needs of people in today's world. It is quite difficult to preserve the essence of modern society without energy.[14] It would not be wrong to say that one of the most important tools determining the quality of human

---

[10] Energy exists in nature independently of human beings. The existence of the universe and its transformation in the historical process have been realized with energy. As human beings discover the world they live in, they have begun to discover the energy and energy sources found in nature. The ability of the wave or wind in water resources to move an object from one place to another is the information that people have obtained by observing and experiencing. In the light of this information, mankind, who discovered the energy sources in nature, started to use wave and wind energy.

[11] Cara New Daggett, *The Birth of Energy*, 15.

[12] Nebojsa Nakicenovic, "The Automobile Road to Technological Change: Diffusion of the Automobile as a Process of Technological Substitution," *Technological Forecasting and Social Change*, 29/4 (1986): 309-340.

[13] Eric Morris, "From Horse Power to Horsepower," *Access Magazine*, 1/30 (2007), 2-10.

[14] Ibrahim Dincer, "Renewable Energy and Sustainable Development: A Crucial Review," *Renewable and Sustainable Energy Reviews* 4/2 (2000): 157-175, 157.

life in the 21st century is energy.[15] People are constantly using energy to improve their living standards.[16] One of the instruments necessary for modern societies to continue their lives in a comfortable and healthy way is considered as energy. In other words, there is an increase in the right direction between the quality of individuals' living standards and energy use/consumption.[17]

In this period we live in, the usage area of energy is quite wide. Energy has an important place in all phases of human life. First, energy has a strategic role in economic growth.[18] It is widely accepted that there is a direct relationship between economic development and energy consumption.[19] Fossil fuels, in particular, have played a critical role in economic development at the global level since the Industrial Revolution.[20] Economic development cannot be achieved without energy.[21] Because all processes such as the production, processing, and transportation of materials needed for economic growth are realized with energy.[22] For example, all technological machines in factories where cars, airplanes, and all other vehicles are mass-produced work with energy. Therefore, energy constitutes the locomotive of economic development.[23]

---

[15] Stanley R. Bull, "Renewable Energy Today and Tomorrow," *Proceedings of the IEEE* 89/8 (2001): 1216-1226.

[16] Mohammad S. Masnadi et al., "From Fossil Fuels Towards Renewables: Inhibitory and Catalytic Effects on Carbon Thermochemical Conversion During Co-Gasification of Biomass with Fossil Fuels," *Applied Energy* 140 (2015): 196-209, 196.

[17] Donald L. Klass, Biomass for Renewable Energy, Fuels, and Chemicals, (London: Elsevier, 1998), 1.

[18] Ather Hassan et al, "Monetization of the Environmental Damage Caused by Fossil Fuels," *Environmental Science and Pollution Research* 28/17 (2021): 21204-21211, 21204.

[19] Alicia Valero and Antonio Valero, "What are the Clean Reserves of Fossil Fuels?," *Resources, Conservation and Recycling* 68 (2012): 126-131, 130.

[20] Shahriar Shafiee and Erkan Topal, "An Econometrics View of Worldwide Fossil Fuel Consumption and the Role of US." *Energy Policy* 36/2 (2008): 775-786.

[21] Frano Barbir, T. Nejat Veziroğlu and H. J. Plass, "Environmental Damage due to Fossil Fuels Use," *International Journal of Hydrogen Energy* 15/10 (1990): 739-749.

[22] Raul A. Barreto, "Fossil Fuels, Alternative Energy and Economic Growth," *Economic Modelling* 75 (2018): 196-220, 196.

[23] Mikael Höök and Xu Tang, "Depletion of Fossil Fuels and Anthropogenic Climate Change—A Review," *Energy Policy* 52 (2013): 797-809, 804.

One of the second most common uses of energy in modern societies is the transportation sector.[24] Approximately 30% of the world's energy consumption is used in the transportation sector.[25] People are constantly moving from one place to another. Especially in big metropolises such as Istanbul, London, and Tokyo, this mobility is much higher than in rural areas. Mobility does not occur only in the city. It can be done from one city to another city or from one country to another. This mobility can be realized for different purposes such as work, education, social life, entertainment, and vacation. It is very difficult to provide mobility between places where the physical distance is high with muscle power. Due to reasons such as timesaving, security, and comfort, mobility between physical spaces is provided by vehicles such as cars, buses, ferries, metro, trams, airplanes, helicopters, and taxis. These means of transportation are produced by machines working with energy. In addition, vehicles work with the use of energy sources. Oil is the most common energy source used in vehicles.[26] In short, energy is used extensively in the transportation sector.[27, 28]

Another common usage area of energy is the communication sector. The century we live in is defined as the communication age.[29] People in different physical places are constantly communicating with each other for reasons such as work, entertainment, and shopping. Communication takes place with the use of technological tools. For example, a person living in Turkey needs technological products such as a telephone, tablet, computer, and internet to communicate with someone

---

[24] Raymond Murray and Keith E. Holbert, Nuclear Energy: An İntroduction to the Concepts, Systems, and Applications of Nuclear Processes (Waltham, Oxford: Elsevier, 2014), 11.

[25] Yuanfeng Wang et al., "Transport Energy Consumption and Saving in China," *Renewable and Sustainable Energy Reviews*, 29 (2014): 641-655, 641.

[26] Yi-Xuan Gao et al., "Road Transport Energy Consumption in the G7 and BRICS: 1973-2010," *International Journal of Global Energy Issues*, 38/4-6 (2015): 342-356.

[27] Ramakrishnan Ramanathan, "A Holistic Approach to Compare Energy Efficiencies of Different Transport Modes," *Energy Policy*, 28/11 (2000): 743-747.

[28] Hinrich Helms and Udo Lambrecht, "The Potential Contribution of Light-Weighting to Reduce Transport Energy Consumption," *International Journal of Life Cycle Assessment*, 12/1 (2007): 58-64.

[29] İbrahim Ortaş, "Bilgi ve İletişim Çağında Bilimsel Bilgiye Erişimin Önemi ve Türkiye'nin Bilgiye Erişim Potansiyeli," *Türk Kütüphaneciliği*, 32/3 (2018): 223-232.

living in Japan verbally, in writing or visually, and in a short time. However, it is not possible to produce and use all these communication tools without energy. Because all the tools necessary for communication are produced and work with energy. For this reason, one of the areas where energy is used the most is the communication sector.[30]

Energy is also widely used in bringing basic needs[31] and luxury goods to the market. In order for people to continue their lives in a healthy way, they must regularly meet their physiological needs such as food, water, sleep, and breathing.[32] At the same time, housing and similar structures are needed in order to protect themselves against physical conditions, that is, to meet their security needs. However, the people of the modern era are not only in an effort to meet their basic needs. In order to further improve living standards and have a more comfortable life, goods and services are purchased other than their basic needs. Technological devices such as individual vehicles, yachts, helicopters, televisions, computers, and dishwashers are included in this scope.

Energy is also used extensively in processes such as the production, processing, and transportation of goods that people consume to improve their lives.[33] For example, agricultural machinery is used in processes such as fertilizing, planting, and harvesting wheat, which is the main raw material of bread, which is indispensable for many families' meals in Turkey. Turning the wheat into flour, transporting the flour to the bakers and (or) buyers, transforming the flour into bread in the bakery, and distribution of the bread to places such as markets are realized by using

---

[30] Gokmen Ceribasi & Muhammed Caliskan, "Short-And Long-term Prediction of Energy to be Produced in Hydroelectric Energy Plants of Sakarya Basin in Turkey," *Energy Sources, Part A: Recovery, Utilization, and Environmental Effects*, (2019): 1-16.

[31] Abraham Maslow, also known as Maslow Theory, categorizes people's needs in order of importance. At the first level of Maslow's hierarchy of needs are physiological needs such as food, water, breathing, sex and sleep. In the second step of the hierarchy of needs, there is the need for security that protects the individual's physical and mental health. The third level is the need for belonging and love. The fourth level is the need for respect. The last step is the need for self-actualization.

[32] Beiwen Chen et al., "Basic Psychological Need Satisfaction, Need Frustration, and Need Strength Across Four Cultures," *Motivation and Emotion*, 39/2 (2015): 216-236.

[33] Katherine D. Arbuthnott and Brett Dolter Kahneman, "Escalation of Commitment to Fossil Fuels," *Ecological Economics* 89 (2013): 7-13.

energy-operated vehicles. Mechanical agricultural activities are dependent on energy.[34] For this reason, the production and delivery of basic necessities and luxury products that we consume occur with energy consumption. The basic needs of the rapidly increasing world population and the luxury consumption goods they use are met in a stable manner thanks to the mass production of energy-operated machines. Otherwise, it is not possible to produce goods and services that can meet the needs of 8 billion people with manual production. Therefore, energy has an important role in reducing poverty.[35]

Energy has a key role in faster and more comfortable access to information.[36] The 21st century we live in is also defined as the Information/Information Age.[37] Technological products are widely used in individuals' access to information. The distance education process, which has become widespread especially with the COVID-19 process[38], has brought the importance of technology in accessing information to the agenda. It is not possible to produce and use technological tools without energy. All technological products such as televisions, radios, computers, telephones, and tablets that enable people to access information are produced with energy. At the same time, these technological tools used in accessing information work with energy. For example, a computer or telephone cannot be operated for a long time without electricity. For this reason, energy enables us to access information more quickly and comfortably.

Finally, energy consumption is very common in meeting the basic

---

[34] Robert Foster, Majid Ghassemi and Alma Cota, *Solar Energy: Renewable Energy and the Environment* (Boca Raton: CRC Press, 2009).

[35] Selcuk Bilgen, Kamil Kaygusuz and Ahmet Sari, "Renewable Energy for a Clean and Sustainable Future," *Energy Sources* 26/12 (2004): 1119-1129, 1128.

[36] Martin Kaltschmitt, Wolfgang Streicher and Andreas Wiese (eds.), *Renewable Energy: Technology, Economics and Environment*, (New York: Springer Science & Business Media, 2007).

[37] Manuel Castells, "An Introduction to the Information Age," *City* 2/7 (1997): 6-16.

[38] Deepika Nambiar, "The Impact of Online Learning During COVID-19: Students' and Teachers' Perspective," *The International Journal of Indian Psychology* 8/2 (2020): 783-793.

needs of people such as heating[39,40] and cooking[41] in residences, the service sector, work areas, and public buildings. In order for people to survive, they need to eat nutrients at certain intervals. Otherwise, it is difficult for them to continue their lives in a healthy way. Energy sources are used in cooking the meals that a person consumes on average 2-3 times a day. Energy is also used to provide heating, which is another important need, especially in the winter season. Generally, people living in rural areas use biomass energy[42] and coal for their cooking and heating needs; Urban residents living in metropolises use natural gas more widely.

3. ENERGY SOURCES

As stated before energy is obtained from different sources. Energy resources are categorized under three main groups. The first group is fossil fuels. Coal, oil, and natural gas are defined as fossil fuels.[43, 44] Fossil fuels are non-renewable and are found in nature in limited quantities. The second main energy source is renewable energy which includes solar energy, wave energy, thermal energy, and wind energy. The third main energy source is nuclear energy. Nuclear energy, which has been used since the second half of the 20th century, is obtained from uranium in nuclear power plants.

3.1. Fossil Fuels

Fossil fuels consisting of organic compounds are also defined as mineral fuels. Fossil fuels are formed by the dissolution of animal and

---

[39] Olivia Guerra, Santin Laure and Itard Henk Vissche, "The Effect of Occupancy and Building Characteristics on Energy Use for Space and Water Heating in Dutch Residential Stock," *Energy and Buildings*, 41/11 (2009), 1223-1232.

[40] Radu Zmeureanu and Guillaume Renaud, "Estimation of Potential Impact of Climate Change on the Heating Energy Use of Existing Houses," *Energy Policy*, 36/1 (2008): 303-310.

[41] Tiffany J. Hager and Ruben Morawicki, "Energy Consumption During Cooking in the Residential Sector of Developed Nations: A Review," *Food Policy* 40 (2013): 54-63.

[42] Vijay Laxmi Pandey and Aditi Chaubal, "Comprehending Household Cooking Energy Choice in Rural India," *Biomass and Bioenergy*, 3511 (2011): 4724-4731, 4724.

[43] Shahriar Shafiee and Erkan Topal, "An Econometrics View," 775.

[44] Guy Ourisson, Pierre Albrecht and Michel Rohmer. "The Microbial Origin of Fossil Fuels," *Scientific American* 251/2 (1984): 44-51, 44.

plant remains buried underground as a result of natural cycles.[45] The formation of fossil fuels, which is the result of the thermochemical reaction of animal and plant wastes in the lower layers of the earth in a certain region, can take millions of years. The formation of fossil fuels can take up to 650 million years.[46] Fossil fuels are energy sources containing high carbon and hydrocarbons.

The discovery and use of fossil fuels, which is the most widely used energy source in the world, from the Industrial Revolution to the present, dates back to ancient times in human history. Fossil fuels have been used by humans for different purposes for hundreds of years. For example, the use of coal for heating and cooking in England in the Middle Ages was common throughout the country.[47] It is still widely used in the world today. Fossil fuels constitute the most important driving force of the industrialized world after the Industrial Revolution and its economic growth.[48] It is estimated that fossil fuels will have the highest share among energy sources in world energy production until at least 2050.[49]

Fossil fuels are non-renewable energy sources.[50] As mentioned before, energy resources are divided into two basic groups as renewable and non-renewable. Fossil fuels are among the non-renewable energy sources. Fossil fuels are in limited quantity in nature.[51] Fossil fuels are energy sources that decrease as they are used and cannot be renewed immediately in case of depletion, that is, there is a risk of extinction.[52] Since the rate of use of existing fossil fuel reserves is higher than the rate of formation of new fossil fuel resources, there is a risk of depletion.[53]

---

[45] Cara New Daggett, *The Birth of Energy*, 25.

[46] Nivedita Patel, "Environmental and Economical Effects of Fossil Fuels," *Journal of Recent Research in Engineering and Technology* 1 (2014): 1-12, 1.

[47] Michael Le Page, "World Without Fossil Fuels," *New Scientist*, 224/2991 (2014): 34-39, 34.

[48] Mikael Höök and Xu Tang, "Depletion of Fossil Fuels," 799.

[49] Paul E. Brockway et al., "Estimation of Global Final Stage Energy Return on Investment for Fossil Fuels with Comparison to Renewable Energy Sources," *Nature Energy* 4/7 (2019): 612-621, 612.

[50] Raul A. Barreto, "Fossil Fuels, Alternative," 196.

[51] Luca Chiari and Antonio Zecca, "Constraints of Fossil Fuels Depletion on Global Warming Projections," *Energy Policy* 39/9 (2011): 5026-5034, 5026.

[52] Raul A. Barreto, "Fossil Fuels, Alternative," 196.

[53] Nivedita Patel, "Environmental and Economical Effects," 1.

Because fossil fuels emerge as a result of natural reactions lasting millions of years, which means it seems not possible for new reserves to be formed in the short and medium-term.

Fossil fuels are considered as a relatively cheap energy source.[54, 55] One of the most important reasons underlying fossil fuel's being the main energy source all over the world for hundreds of years is that it has a more affordable price compared to other energy sources. Fossil fuels are cheaper to extract, process, and transport than renewable energy sources. However, considering the price of fossil fuels only with the cost of processing may lead us to the wrong conclusion. Because fossil fuels have environmental costs.[56] In other words, the cost of the damage caused by fossil fuels to the environment is quite high compared to other energy sources.

One of the most distinctive features of fossil fuels is that they cause environmental pollution.[57] Fossil fuels are considered to be the main cause of environmental problems on a global scale. Because these fuels contain high rates of carbon and hydrogen.[58] With the burning of fossil fuels in residences, industrial facilities, and vehicles, greenhouse gases are released into the atmosphere.[59, 60] To explain with a more concrete example, the combustion of only 1 liter of diesel causes the emission of 2.9 kilograms of greenhouse gases.[61]

Greenhouse gases are pollutants for the physical and biological elements of the environment. The most common greenhouse gases are as follows; carbon dioxide (CO_2), hydrofluorocarbons (HFCs), methane (CH_4), nitrous oxide (N_2O), Sulfur hexafluoride (SF6), and water vapor

---

[54] Nivedita Patel, "Environmental and Economical Effects," 6.

[55] Mason Inman, The True Cost of Fossil Fuels, *Scientific American* 308/4 (2013): 58-61, 60.

[56] Nivedita Patel, "Environmental and Economical Effects."

[57] Cees Withagen, "Pollution and Exhaustibility of Fossil Fuels," *Resource and Energy Economics* 16/3 (1994): 235-242, 235.

[58] Ather Hassan et al., "Monetization of the Environmental," 21204.

[59] Frano Barbir, T. Nejat Veziroğlu and H. J. Plass, "Environmental Damage," 739.

[60] Luca Chiari and Antonio Zecca, "Constraints of Fossil Fuels," 5026.

[61] Ümit Ağbulut and Suat Sarıdemir, "A General View to Converting Fossil Fuels to Cleaner Energy Source by Adding Nanoparticles," *International Journal of Ambient Energy* 42/13 (2021): 1569-1574, 1569.

(H_2O).[62] Carbon dioxide, the most common greenhouse gas in the atmosphere, is responsible for 55% of global warming.[63] The most important cause of carbon dioxide emission is fossil fuels.[64] The emission of greenhouse gases such as carbon dioxide and nitrogen dioxide[65] caused by fossil fuels is one of the main causes of climate change.[66]

The environmental impacts of fossil fuels are not limited to climate change. The use of fossil fuels causes many environmental problems such as air pollution, acid rain, water pollution, soil pollution, ozone depletion, deforestation, and reduction of biodiversity.[67, 68] Fossil fuels also trigger mercury pollution.[69] These environmental problems caused by fossil fuels result in the deterioration of the ecological balance as a whole. Even if greenhouse gas emissions from the use of fossil fuels are cut now, its effects are estimated to last for thousands of years.[70] For this reason, the environmental effects of fossil fuels spread over the long term and threaten future generations.

Environmental pollution caused by the excessive use of fossil fuels poses serious risks to human life.[71] A clean environment is a minimum requirement for people to continue their lives in a healthy way.[72]

---

[62] S. A. Montzka, E. J. Dlugokencky & J. H. Butler, "Non-CO2 Greenhouse Gases and Climate Change," *Nature* 476/7358 (2011): 43-50.
[63] Nada Alrikabi, "Renewable Energy Types," *Journal of Clean Energy Technologies* 2/1 (2014): 61-64, 63.
[64] Mohammad Reza Lotfalipour, Mohammad Ali Falahi and Malihe Ashena, "Economic Growth, CO2 Emissions, and Fossil Fuels Consumption in Iran," *Energy* 35/12 (2010): 5115-5120.
[65] Umair Shahzad, "The Need for Renewable Energy Sources," *Energy* 2 (2012): 16-18, 16.
[66] Mikael Höök and Xu Tang, "Depletion of Fossil Fuels," 800.
[67] Giovanni Nicoletti et al., "A Technical and Environmental Comparison Between Hydrogen and Some Fossil Fuels," *Energy Conversion and Management* 89 (2015): 205-213.
[68] Robert Foster, Majid Ghassemi and Alma Cota, *Solar Energy*, 1.
[69] Oiva I. Joensuu, "Fossil Fuels as a Source of Mercury Pollution," *Science* 172/3987 (1971): 1027-1028.
[70] Toby Tyrrell, John G. Shepherd and Stephanie Castle, "The Long-term Legacy of Fossil Fuels," *Tellus B: Chemical and Physical Meteorology* 59/4 (2007): 664-672, 670.
[71] Ather Hassan et al., "Monetization of the Environmental," 21209.
[72] Julie Kerr Casper, *Fossil Fuels and Pollution: The Future of Air Quality* (New York: Infobase Publishing, 2010).

The quality of the environment directly affects the health of individuals.[73] For this reason, environmental problems caused by fossil fuels cause millions of people to die and get fatal diseases on a global level every year.[74] Because pollutants such as carbon dioxide, sulfur dioxide, and nitrogen oxide, which are produced by the combustion of fossil fuels, cause air pollution, which poses serious risks to human life worldwide.

Fossil fuels are widely used in different fields and sectors to meet human needs. One of the areas where fossil fuels are most widely used is electricity generation.[75] Electricity is one of the important needs used in many phases of our daily lives. It is used in our private and social lives and workplaces. In addition, many technological products are produced and operated with electricity. Electricity is obtained by the conversion of fossil fuels. Coal is the most widely used fossil fuel type in electricity generation in the world. Coal meets approximately 27% of the world's electricity needs.[76]

Another area where fossil fuels are used the most is the transportation sector. In the last century, individual and public vehicles have shown rapid progress all over the world in terms of technology. Due to the rapid increase in the world population, the need for vehicles is increasing, especially in big metropolises such as London, Paris, Bangkok, Tokyo, and Istanbul. Public vehicles such as trains, buses, ferries, and planes, which are widely used by people for different purposes such as work, travel, and socialization, and individual vehicles work with energy.[77] At the same time, energy is used in the production of these

---

[73] Ümit Ağbulut and Suat Sarıdemir, "A General View," 1569.

[74] Kakali Mukhopadhyay and Osmo Forssell, "An Empirical Investigation of Air Pollution from Fossil Fuel Combustion and Its Impact on Health in India During 1973–1974 to 1996–1997," *Ecological Economics* 55/2 (2005): 235-250.

[75] Leif Gustavsson et al., "Climate Change Effects of Forestry and Substitution of Carbon-intensive Materials and Fossil Fuels," *Renewable and Sustainable Energy Reviews* 67 (2017): 612-624, 614.

[76] Marija M. Janković et al., "Study of Radioactivity in Environment around Power Plants Tent a and Kolubara due to Coal Burning for 2015," *RAD Conference Proceedings* 1 (2016): 84-89, 84.

[77] Canan G. Corlu et al., "Optimizing Energy Consumption in Transportation: Literature Review, Insights, and Research Opportunities," *Energies*, 13/5 (2020): 1115-1148.

vehicles. Without energy, it is not possible to meet the transportation needs of billions of people comfortably and safely.

The main energy source used in public and individual vehicles is fossil fuels.[78] The world transportation sector is dependent on fossil fuels. The main energy source used in vehicles is oil. In 2018, 90% of the fuel needs of individual and public vehicles in the world were met from oil.[79] However, it should be noted that the use of electric vehicles is becoming more and more common. It is estimated that 50 million electric vehicles will be sold in 2030.[80] It is possible to say that the share of fossil fuels in the transportation sector will tend to decrease due to the increasing share of renewable energy sources in energy production and the widespread use of electric vehicles.

Another area where fossil fuels are widely used is heating.[81] People need warming to protect themselves from seasonal conditions. Fossil fuels are also used to meet the heating needs of people, especially in winter.[82] There is a need for heating tools in order to be protected from cold weather in the winter season, which is seen in geographical regions with the Mediterranean climate, oceanic climate, continental climate, tundra climate, and arctic climate. The use of coal and natural gas is quite common in these regions. Coal is widely used for heating in developing countries, especially in rural areas.[83] However, many developed countries have restricted the use of coal for heating due to environmental pollution.[84]

---

78 Elisabeth Ekener-Petersen, Jonas Höglund and Göran Finnveden, "Screening Potential Social Impacts of Fossil Fuels and Biofuels for Vehicles," *Energy Policy* 73 (2014): 416-426, 418.

79 "Transport," last updated date: 17 Şubat, 2022 https://www.bp.com/en/global/corporate /energy-economics/energy-outlook/demand-by-sector/transport.html

80 "World Energy Outlook," last updated date: 12 October, 2021 https://iea.blob.core.windows.net/assets/a72d8abf-de08-4385-8711-b8a062d6124a/WEO2020.pdf, 138.

81 Lucy Allen et al., "Fossil Fuels and Water Quality," İçinde *the World's Water*, ed. Peter H. Gleick (Washington, DC: Island Press, 2012), 73.

82 Nivedita Patel, "Environmental and Economical Effects."

83 Ying Zhou et al., "Impact of Rural Residential Coal Combustion on Air Pollution in Shandong, China," *Chemosphere* 260 (2020): 127517.

84 Aiymgul Kerimray et al., "Coal Use for Residential Heating: Patterns, Health Implications and Lessons Learned," *Energy for Sustainable Development*, 40 (2017): 19-30, 20.

Natural gas is used as a means of heating in developed countries and large metropolises.[85]

Fossil fuels are also a widely used energy source in cooking.[86] People started to cook with the discovery of fire. In ancient times, biomass energy was used in cooking food. With the discovery of fossil fuels, the use of coal began to come to the fore. After the Industrial Revolution, one of the most used energy sources in cooking was coal.[87] Today, fossil fuels are still preferred for cooking. However, this situation differs according to the economic and demographic structures of the countries. The use of coal in cooking is common in traditional societies, as it is easy to access and more affordable than other energy sources. For example, one of the most used energy sources for heating and cooking in rural areas of China is coal.[88]

Fossil fuels are also used in the field of industrial agriculture.[89] The history of agricultural activities dates back to ancient times. People have been engaged in agricultural activities for more than ten thousand years. However, the use of fossil fuels in agricultural activities and industrial agriculture started in the second half of the 20th century.[90] One

---

[85] Xiaoli Duan et al., "Household Fuel Use for Cooking and Heating in Chin`a: Results from the First Chinese Environmental Exposure-Related Human Activity Patterns Survey (CEERHAPS)," *Applied Energy* 136 (2014): 692-703.

[86] By burning coal in a heating device called "soba" in rural areas in Turkey, both heating needs are met and food is cooked. With the uncontrolled urbanization process that gained momentum between the years 1950-1980, this situation in rural areas was also evident in big metropolises such as Istanbul, Ankara and Izmir. With the widespread use of natural gas, the use of coal as a heating and cooking tool has decreased.

[87] With the widespread use of natural gas, the use of coal as a heating and cooking tool has decreased. The role assigned to women in traditional societies is to do housework. Cooking and using the heating device to heat the house are commonly done by women. Women, who spend more time at home than men, are more exposed to air pollution caused by the use of coal, and as a result, they suffer from more health problems.

[88] Wah Kit Lam, "Lung Cancer in Asian Women—The Environment and Genes," *Respirology*, 10/4 (2005): 408-417, 411.

[89] Cutler J. Cleveland, "The Direct and Indirect Use of Fossil Fuels and Electricity in USA Agriculture, 1910–1990," *Agriculture, Ecosystems & Environment* 55/2 (1995): 111-121, 111.

[90] Leo Horrigan, Robert S. Lawrence and Polly Walker P., "How Sustainable Agriculture Can Address the Environmental and Human Health Harms of Industrial Agriculture," *Environmental Health Perspectives*, 110/5 (2002): 445-456, 446.

of the most important reasons for this is to meet the nutritional needs of the rapidly increasing world population. In order to get faster and more inputs, machines working mostly with fossil fuels are used in industrial agriculture. Some of these machines are as follows; stone harvesting machine, soil tillage machine, seed planting machine, grain cart, silage making machine, conveyor belt, cotton picker, grain dryer, baler, etc. At the same time, the vehicles used to transport agricultural products from fields to marketplaces and buyers also work with fossil fuels.

3.1.1. Coal

The history of using coal as a fuel dates back to ancient times. Evidence that coal was used as a fuel has been found in China in the 1000s BC.[91] Today, coal is the most widely used energy source in electricity generation.[92] Coal emerges with the transformation of plants over time.[93] Coal takes millions of years to form. It is the longest-lasting energy source among fossil fuels.[94] It is not possible for coal to be regenerated in nature in the short term. For this reason, coal is considered as a non-renewable energy source. 60% of the limited amount of coal in the world is obtained from underground mining.[95]

The use of coal in energy production tends to decrease gradually. The coal-based energy production process, which started with the Industrial Revolution, tends to decrease due to the development of environmental awareness and the increase in the use of alternative energy sources. At the beginning of the 20th century, the share of coal in world energy production is more than 50%.[96] But today, this rate has decreased to 33%. It is estimated that the use of coal will decrease further, especially with the increase in investments in renewable energy sources.

---

[91] Cara New Daggett, *The Birth of Energy*, 27-28.

[92] Tomas Kåberger, "Progress of Renewable Electricity Replacing Fossil Fuels," *Global Energy Interconnection* 1/1 (2018): 48-52, 50.

[93] Guy Ourisson, Pierre Albrecht and Michel Rohmer, "The Microbial Origin," 44.

[94] Cara New Daggett, *The Birth of Energy*, 27.

[95] Bian Zhengfu et al., "Environmental Issues from Coal Mining and Their Solutions," *Mining Science and Technology (China)* 20/2 (2010): 215-223, 216.

[96] Vladimir S. Arutyunov and Georgiy V. Lisichkin, "Energy Resources of the 21st Century: Problems and Forecasts. Can Renewable Energy Sources Replace Fossil Fuels?," *Russian Chemical Reviews* 86/8 (2017): 777-804, 781.

Coal reserves in the world are more than the reserves of other fossil fuel types.[97] As of 2020, the identified exploitable coal reserves in the world are around 1074 billion tons.[98] Coal reserves in the world are located in different geographical regions. There are 6 countries where coal reserves are widely found.[99] As seen in Table 1, the region with the highest coal reserves is the United States of America. 23% of the world's coal reserves are located within the borders of the USA. Russia is in second place. Russia has 15% of the world's coal reserves. Ranking third, Australia has 14% of the world's coal reserves within its borders. Ranking fourth, the People's Republic of China has 13% of the world's coal reserves. 35% of the coal reserves are located in other countries.

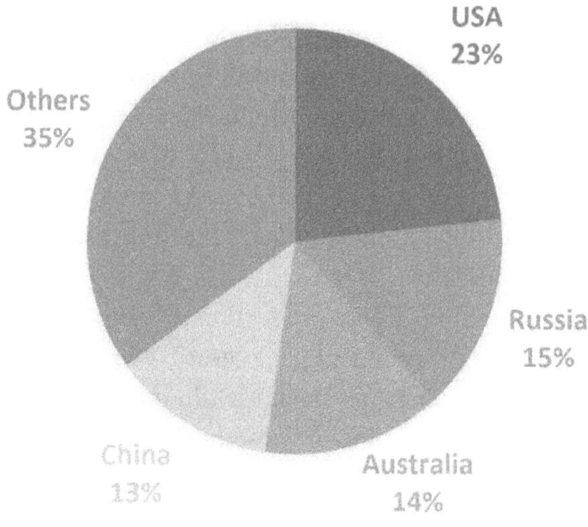

Table 1: Distribution of Coal Reserves by Country in 2020 [100]

Coal is the most polluting energy source among the energy sources.[101] Because the energy source containing the most carbon is

[97] Nivedita Patel, "Environmental and Economical Effects," 1.

[98] "Statistical Review of World Energy 2021," last updated date: 19 October, 2021 https://www.bp.com/content/dam/bp/business-sites/en/global/corporate/pdfs/energy-economics/statistical-review/bp-stats-review-2021-coal.pdf, 47.

[99] Mikael Höök and Xu Tang, "Depletion of Fossil Fuels," 799.

[100] "Statistical Review of World Energy 2021," 47.

[101] D. Yogi Goswami and Frank Kreith eds., *Handbook of Energy Efficiency and Renewable Energy* (Boca Raton: Crc Press, 2007).

coal.[102] With the combustion of coal, greenhouse gases and polluting gases such as carbon dioxide, sulfur dioxide, nitrogen oxide, and chlorofluorocarbons are released into the atmosphere.[103] One of the most important factors causing the emission of greenhouse gases is coal thermal power stations.[104] Fine particulate matter ($PM_{2.5}$), which is the riskiest for human health among air pollutants, comes out with the combustion of coal.[105] The combustion of coal in coal power plants produces bottom ash and fly ash, which cause environmental pollution.[106] At the same time, coal mining causes the emission of methane gas, one of the most harmful greenhouse gases.[107] Accordingly, it has been determined that there is a dangerous level of air and water pollution in the areas where coal is burned.[108] These gases and pollutants also trigger environmental problems such as soil pollution and climate change.

Environmental pollution caused by the use of coal triggers many deadly diseases.[109, 110] The effects of coal on public health have been known for many years. It was first reported by Percival Pott in 1775 that the pollutants and gases produced by the burning of coal triggered the cancer disease.[111] Lung cancer is the most important health problem faced

---

[102] Shannon Elizabeth Bell and Richard York, "Coal, Injustice, and environmental Destruction: Introduction to the Special İssue on Coal and the Environment," *Organization & Environment* 25/4 (2012): 359-367, 359.
[103] Ahmad Shamshad, Fulekar M. H., and Pathak Bhawana, "Impact of Coal Based Thermal Power Plant on Environment and Its Mitigation Measure," *International Research Journal of Environment Sciences* 1/4 (2012): 60-64, 60.
[104] Vishwanath Haily Dalvi, Sudhir V. Panse and Jyeshtharaj B. Joshi, "Solar Thermal Technologies as a Bridge From Fossil Fuels to Renewables," *Nature Climate Change* 5/11 (2015): 1007-1013, 1007.
[105] Wenkai Guo et al., "Ambient $PM_{2.5}$ and Related Health Impacts of Spontaneous Combustion of Coal and Coal Gangue," *Environmental Science & Technology*, 55/9 (2021), 5763-5771.
[106] Marija M. Janković et al., "Study of Radioactivity," 84.
[107] Bian Zhengfu et al., "Environmental Issues," 218.
[108] Sribas Goswami, "Impact of Coal Mining on Environment," *European Researcher* 92/3 (2015):185-196, 186.
[109] Nivedita Patel, "Environmental and Economical Effects," 2.
[110] William M Castleden et al., "The Mining and Burning of Coal: Effects on Health and the Environment," *The Medical Journal of Australia* 195/6 (2011): 333-335.
[111] Joellen Lewtas, "Air Pollution Combustion Emissions: Characterization of Causative Agents and Mechanisms Associated with Cancer, Reproductive, and Cardiovascular Effects," *Mutation Research/Reviews in Mutation Research* 636/1-3 (2007): 95-133, 96.

by those exposed to air pollution caused by the use of coal.[112] For this reason, the Xuanwei region of China, where coal mining and coal use are common, is called a "cancer village."[113]

3.1.2. Oil

Oil is a type of fossil fuel in liquid form. It consists of decomposed animal and plant remains that accumulate on the seafloor.[114] There are two basic types of oil, also called crude oil. The first is light oil with high gravity. Light oil can be green, yellow, or light brown in color. The second type, heavy oil, is of low gravity. Heavy oil is black or dark brown in color.[115]

Although oil has been widely used in recent centuries, the discovery of oil is not a new phenomenon. Primitive peoples discovered oil. In ancient times, oil was used for heating and cooking by burning.[116] Since there was no technological progress, energy production from oil was not seen until the Industrial Revolution. The use of oil as an energy source began in the 1800s.[117] The first oil well in the world was drilled in the USA in 1859.[118] By the 1940s, oil had become the world's most dominant energy source.[119] Today, the importance of oil still continues among the energy sources.

---

[112] H. Dean Hosgood et al., "Household Coal Use and Lung Cancer: Systematic Review and Meta-analysis of Case–Control Studies, with an Emphasis on Geographic Variation," *International journal of Epidemiology*, 40/3 (2011): 719-728.

[113] Mengyuan Zhang et al., "Hemolysis of PM10 on RBCs in Vitro: An Indoor Air Study in a Coal-burning Lung Cancer Epidemic Area," *Geoscience Frontiers* 13/1 (2022): 101176.

[114] Nivedita Patel, "Environmental and Economical Effects," 2.

[115] Hüseyin Naci Bayraç, "Küresel Petrol Piyasasında Fiyat Oluşumu ve Ekonomik Etkileri," *Yalova Sosyal Bilimler Dergisi* 9/19 (2019): 44-59, 46.

[116] Bilgehan Emeklier and Nihal Ergül, "Petrolün Uluslararası İlişkilerdeki Yeri: Jeopolitik Teoriler ve Petropolitik," *Bilge Strateji* 2/3 (2010): 59-85.

[117] Cem Gökçe, "Önemli Bir Enerji Girdisi Olan Petrolün Ekonomik Kalkınma Sürecindeki Rolü," *Afyon Kocatepe Üniversitesi İktisadi ve İdari Bilimler Fakültesi Dergisi* 16/1(2015): 143-154.

[118] Kenneth J. Chew, "The Future of Oil: Unconventional Fossil Fuels," Philosophical Transactions of the Royal Society A: Mathematical, Physical and Engineering Sciences 372/2006 (2014): 1-34, 3.

[119] Volkan Ş. Ediger, "An Integrated Review and Analysis of Multi-energy Transition from Fossil Fuels to Renewables," *Energy Procedia* 156 (2019): 2-6, 3.

Oil has a wide range of uses. In today's world, oil is most commonly used in the transportation sector.[120] It is considered to be the most important energy source because it is difficult or impossible to meet the world's transportation needs without oil with current technology.[121] Oil is used not only as an energy source but also as an input in production.[122] Oil is used as intermediate goods and processed as final goods in the textile industry and defense industry.[123, 124] Finally, oil is also used in electricity generation worldwide.

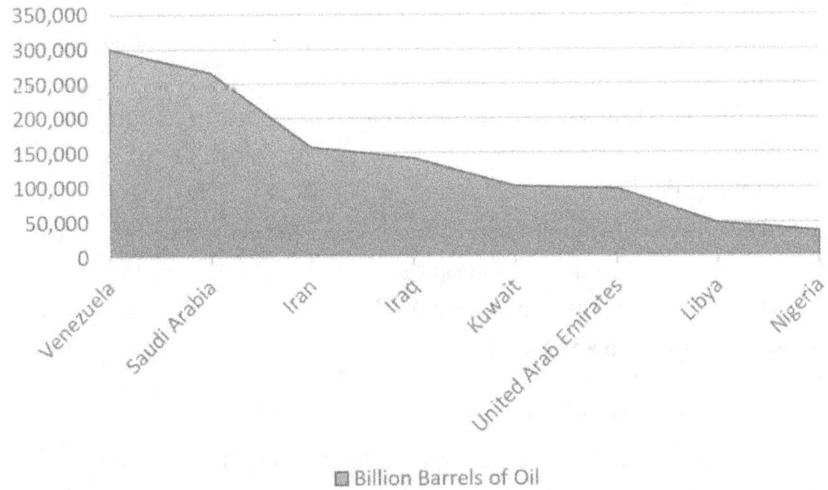

Table 2: Distribution of Oil Reserves By Country in 2020 [125]

As of 2020, the country with the largest oil reserves is Venezuela,

---

[120] Tomas Kåberger, "Progress of Renewable," 49.
[121] Michael Dale, Susan Krumdieck and Pat Bodger, "Net Energy Yield from Production of Conventional Oil," *Energy Policy* 39/11 (2011): 7095-7102, 7095.
[122] Fatma Bahar Şanlı and Kadir Tuna, "Türkiye'de Petrol Tüketimi ile Ekonomik Büyüme Arasındaki İlişkinin Analizi," *Maliye ve Finans Yazıları* 102 (2014): 43-58.
[123] Rebekah Yang, Hasan Ozer and Imad L. Al-Qadi, "Regional Upstream Life-cycle Impacts of Petroleum Products in the United States," *Journal of Cleaner Production* 139 (2016): 1138-1149, 1138.
[124] Mehmet Alagöz, Nur Dilbaz Alacahan and Yağmur Akarsu, "Petrol Fiyatlarının Makro Ekonomi Üzerindeki Etkisi - Ülke Karşılaştırmaları ile Panel Veri Analizi," *Karamanoğlu Mehmetbey Üniversitesi Sosyal ve Ekonomik Araştırmalar Dergisi* 19/33 (2017): 144-150.
[125] "OPEC Annual Statistical Bulletin," last updated date 19 October, 2021, file:///C:/Users/emrah/Downloads/ASB_2020.pdf, 22.

a Latin American country. Venezuela has approximately 300.00 billion barrels of oil. The second country with the largest oil reserves is Saudi Arabia, located in the Middle East region. There are approximately 266.00 billion barrels of oil on the borders of this country. Iran, Iraq, Kuwait, and the United Arab Emirates follow Saudi Arabia in the ranking of the countries with the highest oil reserves. While oil reserves in Iran are around 158.00 billion barrels, this rate is approximately 142.00 billion barrels in Iraq. Oil reserves in Kuwait and the United Arab Emirates are around 101.00 billion barrels and 97 billion barrels, respectively. This information shows that the Middle East region is the geography with the highest oil reserves in the world.

There are also rich oil reserves on the African continent. The countries with the largest oil reserves in Africa are Libya. As of 2020, there are 48,000 billion barrels of oil reserves identified in Libya. Nigeria is the second country with the largest oil reserves in the African continent. The oil reserves identified in Nigeria are around 37,000 billion barrels. Compared to the oil reserves in the Middle East region, it is seen that the oil reserves of Africa are at a very low level.

3.1.3. Natural Gas

Natural gas is an energy source formed by the transformation of organic materials exposed to intense heat over millions of years.[126] Natural gas is formed by the transformation of animal and plant remains accumulated on the sea floor.[127] Natural gas can be found underground freely or together with oil.[128] Natural gas is also considered a derivative of petroleum. Because it consists of hydrocarbons, which is one of the important components of petroleum.

The history of producing energy from natural gas, which can be used in liquid and solid forms, is similar to petroleum. The use of natural gas as an energy source started with technological progress after the Industrial Revolution. As a matter of fact, the first gas well in the world

---

[126] Lucy Allen et al., "Fossil Fuels and Water Quality."

[127] Nivedita Patel, "Environmental and Economical Effects," 2.

[128] Okan Yardımcı, "Türkiye Doğal Gaz Piyasası: Geçmiş 25 Yıl, Gelecek 25 Yıl," *Ekonomi Bilimleri Dergisi*, 3/2 (2011): 157-166, 158.

was drilled in the USA in 1825.[129] Natural gas has become the most important alternative energy source against oil towards the end of the 19th century.[130] Natural gas is widely used in electricity generation and heating in houses.[131] At the same time, natural gas is used in vehicles.[132]

As of 2020, the natural gas reserves that can be determined with the current technology in the world is 181.1 Tcm.[133] Natural gas reserves are mostly found in countries in the Asian continent. The country with the largest natural gas reserves is the Russian Federation. 20.43% of the world's natural gas reserves are located in Russia. The detected natural gas reserve of Russia is 37 Tcm. After Russia, the other country with the highest natural gas reserves is the Islamic Republic of Iran. Iran's 32 Tcm natural gas reserves correspond to 17.6% of the world's natural gas reserves. The third country with the highest natural gas reserves is Qatar. Qatar's 25 Tcm natural gas reserves constitute 13.8% of the world's natural gas reserves.

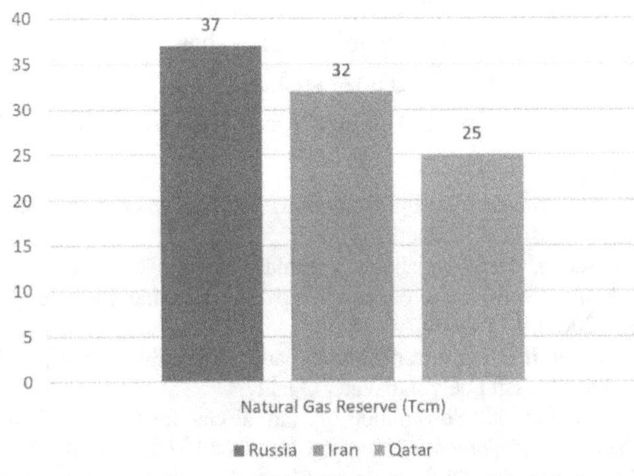

Table 3: Countries with the Most Natural Gas Reserves in 2020 [134]

[129] Kenneth J. Chew, "The Future of Oil," 3.
[130] Volkan Ş. Ediger, "An Integrated Review," 3.
[131] Mahmood Farzaneh-Gord, Shahram Hashemi and Meisam Sadi, "Energy Destruction in Iran's Natural Gas Pipe Line Network," *Energy Exploration & Exploitation* 25/6 (2007): 393-406.
[132] Muhammad Imran Khan, "Policy Options for the Sustainable Development of Natural Gas as Transportation Fuel," *Energy Policy* 110 (2017): 126-136.
[133] It is possible to find more reserves with the developing technology in this area.
[134] "Statistical Review of World Energy 2021," 35.

Natural gas is the least harmful energy source among fossil fuels.[135] It is less harmful to the environment compared to oil and coal.[136] As a matter of fact, natural gas is preferred more than other fossil fuels due to its low environmental impact.[137] However, it should be noted that natural gas, like coal and oil, causes greenhouse gas emissions when burned.[138, 139] The use of natural gas triggers the release of polluting gases such as methane, sulfur dioxide, and black carbon that pollute the atmosphere. The emissions of pollutants caused by the use of natural gas are at lower levels compared to other fossil fuels.[140] For this reason, it is determined that air pollution decreases in regions where natural gas is used instead of coal.[141]

3.2. Renewable Energy

As can be understood from the name of renewable energy, it is the energy that the source used to produce energy constantly renews itself.[142] Renewable energy is the energy obtained from the natural cycles of the sun and the universe.[143] Renewable energy has a key role in ensuring sustainable development.[144] Because sustainable development requires the stable use of energy resources over a long period of time.[145] Due to the

---

[135] David Bodansky, *Nuclear Energy: Principles, Practices, and Prospects*, (New York: Springer Science & Business Media, 2007), 7.

[136] Stefano Di Pascolia, Aldo Femiab and Tommaso Luzzat, "Natural Gas, Cars and the Environment. A (relatively) 'Clean' and Cheap Fuel Looking for Users," *Ecological Economics* 38/2 (2001): 179-189, 179.

[137] Selcuk Bilgen, Kamil Kaygusuz and Ahmet Sari, "Renewable Energy," 1120.

[138] Lucy Allen et al., "Fossil Fuels and Water Quality."

[139] Katharine Hayhoe et al., "Substitution of Natural Gas for Coal: Climatic Effects of Utility Sector Emissions," *Climatic Change* 54/1 (2002): 107-139.

[140] Margarita M. Balmaceda, "Differentiation, Materiality, and Power: Towards a Political Economy of Fossil Fuels," *Energy Research & Social Science* 39 (2018): 130-140, 134.

[141] Murat Çetin and Oğuz Kürşat Demirci, "Erzincan'da Doğal Gaz Kullanımının Hava Kalitesine Etkisi," *Erzincan Üniversitesi Fen Bilimleri Enstitüsü Dergisi* 9/1 (2016): 8-18.

[142] Stanley R. Bull, "Renewable Energy Today," 1216.

[143] Omar Ellabban, Haitham Abu-Rub and Frede Blaabjerg, "Renewable Energy Resources: Current Status, Future Prospects and Their Enabling Technology," *Renewable and Sustainable Energy Reviews* 39 (2014): 748-764, 749.

[144] Atte Harjanne and Janne M. Korhonen, "Abandoning the Concept of Renewable Energy," *Energy Policy* 127 (2019): 330-340.

[145] The concept of sustainability has multiple dimensions. With its most common definition, sustainability is the use of natural resources of the current generation at a level

high risk of extinction of fossil fuels, renewable energy sources have strategic importance in sustainable development[146]. As long-term energy security is possible with renewable energy sources[147], renewable energy is also defined as "*sustainable energy*."[148]

On the other hand, it should be noted that the source of renewable energy is not infinite. There is a risk of depletion or extinction of the source of renewable energy. For example, the source of solar energy, which is one of the most widely used renewable energy sources, is the sun. It is estimated that the sun, which is about 5 billion years old, will go out in 4-5 billion years.[149] In addition, the technological tools used to produce energy from renewable energy sources are made of materials that are found in nature in limited quantities and are at risk of depletion.[150] For these reasons, it would not be wrong to state that renewable energy sources also have a limit, that is, the risk of "non-renewal."

Renewable energy sources meet approximately 10% of the world's energy needs. From this point of view, renewable energy sources are not a substitute for fossil energy sources, but a supplementary energy source.[151] With the current technology, renewable energy sources are not capable of replacing fossil energy sources.[152] However, it should be noted that since the technology in the field of renewable energy is in the

---

that will allow future generations to continue their lives. It is the fact that the resources used to increase the welfare level of people do not harm the physical and biological elements of the environment. For example, fresh water resources are limited in the world. If this resource is consumed or polluted by the current generation, future generations will not be able to meet their water needs and continue to live. In order to sustain life, resources must be used with consideration for future generations and the environment must be protected as a whole.

[146] Ibrahim Dincer, "Renewable Energy," 174.

[147] Nada Alrikabi, "Renewable Energy Types."

[148] Robert Foster, Majid Ghassemi and Alma Cota, *Solar Energy*, 1.

[149] Ted Nield, Supercontinent: Ten Billion Years in the Life of Our Planet (Cambridge: Harvard University Press, 2007).

[150] Atte Harjanne and Janne M. Korhonen, "Abandoning the Concept," 336.

[151] Richard Green and Iain Staffell, "Electricity in Europe: Exiting Fossil Fuels?," *Oxford Review of Economic Policy* 32/2 (2016): 282-303.

[152] Richard York and Shannon Elizabeth Bell, "Energy Transitions or Additions?: Why a Transition from Fossil Fuels Requires more than the Growth of Renewable Energy," *Energy Research & Social Science* 51 (2019): 40-43, 40.

development stage,[153] it is accepted that its share in energy production will increase in the coming years and it can be an energy source that can replace fossil fuels. For example, while 30 kilowatt-hours of energy were produced from wind turbines with a rotor diameter of 10 meters in the 1970s, wind turbines with a rotor diameter of 66 meters were produced in 1997, producing 1,650 kilowatt-hours of energy.[154] It is estimated that two-thirds of the world's energy needs will be met by renewable energy sources by 2050 since the technology in the field of renewable energy is in the development stage.[155]

Quantitative determination of the world's renewable energy resources potential is difficult. Because with the developing technology in this field, the efficiency obtained from renewable energy sources is also increasing. The accepted information is that renewable energy sources have the potential to meet the energy needs of the world. For example, it is possible to obtain more than 1000 times the energy obtained from fossil fuels consumed in the world so far from the sun's rays reaching the earth's surface.[156]

Renewable energy sources have both advantages and disadvantages for the environment and people. One of the advantages of renewable energy sources is diversity.[157] Renewable energy sources are not a type of energy obtained from a single source.[158] Nuclear energy is obtained from uranium sources, fossil energy is obtained from oil, coal, and natural gas sources. In other words, the resources of these two types of energy are quite limited. However, the diversity of renewable energy in

---

[153] The most important development in renewable energy sources is the development of energy storage technology. Renewable energy sources are obtained by converting kinetic energy. Therefore, there is a need for an advanced energy storage system in order to store the converted energy and use it in the future. Although wind and solar energy have been used by people for a thousand years, serious progress in technology in the field of renewable energy has been experienced in the last 40-50 years.

[154] Wim C. Turkenburg, "Renewable Energy Technologies," içinde *World Energy Assessment, Energy and the Challenge of Sustainability* (New York: UNDP, 2000), 232.

[155] Dolf Gielen et al., "The Role of Renewable Energy in the Global Energy Transformation," *Energy Strategy Reviews* 24 (2019): 38-50, 38.

[156] Stanley R. Bull, "Renewable Energy Today," 1216.

[157] Robert Gross, Matthew Leach and Ausilio Bauen, "Progress in Renewable Energy," *Environment International* 29/1 (2003): 105-122, 105.

[158] Atte Harjanne and Janne M. Korhonen, "Abandoning the Concept," 334.

terms of technology and resources has a much richer structure compared to other types of energy sources. It has a wide range of renewable energy sources such as solar energy, wind energy, wave energy, biomass energy, hydroelectric energy, and thermal energy.[159]

Second, the resource used in generating energy in renewable energy is free.[160] For example, there is no cost for the sun used to generate energy from solar energy or the wind used to generate energy with wind energy. In addition, everyone has the opportunity to access these resources. Sun and wind are two important resources that exist in all physical spaces of nature. In fossil and nuclear energy, resources are not free. It is possible to reach resources such as coal, oil, natural gas, and uranium only if a certain cost is paid. In addition, these resources are not accessible to everyone. They are resources that only states, and state-approved institutions and organizations can access. From this point of view, it is possible to say that the cost of resources used in renewable energy is much more affordable than other energy resources.

Third, renewable energy sources are defined as clean energy.[161] Among the energy sources, renewable energy has the least environmental risk. While generating energy from renewable energy sources, there is no direct emission of greenhouse gases or pollutants. In addition, unlike nuclear energy, it does not have risks such as hazardous waste and radioactive contamination in case of an accident. For this reason, renewable energy sources are also called green energy and alternative energy sources[162] to fossil fuels.

The fourth main advantage of renewable energy sources is inclusivity. Fossil fuels are energy sources found in certain parts of the world. For example, 81% of the world's oil reserves are in only eight

---

[159] Nada Alrikabi, "Renewable Energy Types."

[160] Itziar Lazkano, Linda Nøstbakken and Martino Pelli, "From Fossil Fuels to Renewables: The Role of Electricity Storage," *European Economic Review* 99 (2017): 113-129, 115.

[161] N. L. Panwar, S. C. Kaushik and Surendra Kothari "Role of Renewable Energy Sources in Environmental Protection: A Review," *Renewable and Sustainable Energy Reviews* 15/3 (2011): 1513-1524.

[162] Ayhan Demirbaş, "Global Renewable Energy Resources," *Energy Sources* 28/8 (2006): 779-792, 779.

countries, while 70% of the natural gas reserves are in six countries.[163] But renewable energy sources exist in different forms in every region of the world.[164] Resources such as sun, wind, wave, biomass, and geothermal are seen in every region. However, it should be noted that it is not possible to obtain equal efficiency from renewable energy sources in the whole geography.[165] Each geography has different climatic conditions and geographical features. Due to these differences, the solar energy potential of some regions may be higher than other regions.[166] For this reason, it is not possible to get the same efficiency from renewable energy sources in the whole geography.

Renewable energy sources also have many disadvantages. These can be grouped under four headings. The first disadvantage of renewable energy sources is instability. It is not possible to get the same efficiency from renewable energy sources all the time. Efficiency from renewable energy sources can differ from hour to hour, day to day, and month to month. Because renewable energy sources are directly affected by climatic conditions. For example, the amount of energy that can be obtained from solar energy at noon may differ significantly compared to evening time. Similarly, the amount of energy to be obtained from wind energy in winter may be higher than the amount of energy that can be obtained in summer. For this reason, it is not possible to obtain the same amount of energy from renewable energy in a stable manner with current technology.

The second main disadvantage of renewable energy sources is that it triggers visual pollution.[167], which is an unsightly disorder in the artificial environment due to the inappropriate selection of physical elements in terms of aesthetics and functionality.[168] Some of the renewable energy sources necessitate the use of large areas. For example, solar panels

---

[163] Selcuk Bilgen, Kamil Kaygusuz and Ahmet Sari, "Renewable Energy," 1119.

[164] Robert Foster, Majid Ghassemi and Alma Cota, *Solar Energy*, 4.

[165] Alchris Woo Go et al., "Potentials of Agricultural and Agro-industrial Crop Residues for the Displacement of Fossil Fuels: A Philippine Context," *Energy Strategy Reviews* 23 (2019): 100-113.

[166] Robert Gross, Matthew Leach and Ausilio Bauen, "Progress in Renewable Energy," 106.

[167] Robert Foster, Majid Ghassemi and Alma Cota, *Solar Energy*, 4.

[168] Tülün Özer Paşalıoğlu, "Kentsel Alanlarda Görsel Kirlilik: Tekirdağ Örneği," (Master's thesis, Trakya Üniversitesi Fen Bilimleri Enstitüsü, 2007).

and wind turbines are built in large areas within the natural and artificial environment.[169] The fact that these energy sources are not in harmony with the physical elements in the artificial and natural environment causes a visually unsightly appearance.

Renewable energy sources also trigger artificial environmental pollution. Artificial environmental pollution can be defined as the destruction or deterioration of the habitats of other living species by uncontrolled conversion of the natural environment to an artificial environment for human needs.[170] The absence of sufficient physical spaces in the artificial environment for renewable energy sources necessitates spreading towards the natural environment. Conversion of the natural environment, which is the center of biodiversity, for renewable energy sources causes the destruction of the habitats of plant and animal species, in other words, environmental pollution.

Finally, renewable energy sources cause depletion of natural resources. The energy used for the needs of people not only causes the depletion of energy resources, but also other natural resources.[171] The tools required to produce energy are not only energy sources. Many natural resources must be used in the conversion, storage, and transportation of energy resources. For example, the following metals are needed in the production of solar panels used for solar energy; silver, aluminum, copper, indium, cadmium, silica, and gallium.[172] These and similar materials, which are found in limited quantities in nature, are at risk of extinction due to the increase in energy use.

3.2.1. Solar Power

Solar energy is the energy obtained from the sun. It is the energy produced by the conversion of sunlight into energy. The planet we live in

---

[169] Wim C. Turkenburg, "Renewable Energy Technologies," 247.

[170] Emrah Akyüz, Çevre Biliminin ABC'si: Yeni Başlayanlar İçin Çevre Sorunları ve Politikaları (Ankara: Seçkin Yayıncılık, 2020).

[171] Jamie Speirs, Christophe McGlade and Raphael Slade, "Uncertainty in the Availability of Natural Resources: Fossil Fuels, Critical Metals and Biomass," *Energy Policy* 87 (2015): 654-664.

[172] Damien Giurco et al., "Requirements for Minerals and Metals for 100% Renewable Scenarios," İçinde *Achieving the Paris Climate Agreement Goals*, ed. Sven Teske (Cham: Springer, Cham, 2019).

has a very rich potential in terms of solar energy. The world is in the position of a collector that stores a large amount of solar energy. The energy that can be obtained from the sunlight coming to the world in 1 hour is more than the energy need of the world for 1 year.[173] More than 5000 thousand times the energy consumed in the world can be obtained from solar energy.[174] To explain with a more concrete example, 10^7 joules of energy released by the 7.8-magnitude earthquake that occurred in San Francisco in 1906 is equal to the energy released by the sun in one second on Earth.[175]

Solar energy is divided into two basic groups as active solar energy and passive solar energy.[176] Active solar energy is the conversion of sun rays into energy using technological devices. Solar panels installed on the roofs of houses to provide electricity or hot water produced using solar panels in solar fields are considered within this scope. Solar cells used in active solar energy systems convert solar radiation into energy.[177] Passive solar energy, on the other hand, is the direct or indirect use of the sun's rays for human needs without using any intermediary tools. For example, the sun rays entering from the windows of our house contribute to the increase in the temperature of the house, that is, to the heating, can be given as an example of passive solar energy. Passive solar energy has been used since the beginning of human history.

There are two basic forms of solar energy. These are solar thermal energy and solar photovoltaic energy.[178] Thermal solar energy is the conversion of heat from the sun into energy by means of intermediary technological products. Thermal solar energy is generally used to supply hot water in residences, hospitals, factories, public buildings, and hotels.[179]

---

[173] Robert Foster, Majid Ghassemi and Alma Cota, *Solar Energy*, 4.
[174] Daniel M. Kammen, "The Rise of Renewable Energy," *Scientific American* 295/3 (2006): 84-93, 86.
[175] George W. Crabtree and Nathan S. Lewis, "Solar Energy Conversion," *Physics Today*, 60/3 (2007): 37-42, 37.
[176] Nada Alrikabi, "Renewable Energy Types," 62.
[177] Claes G. Granqvist, "Solar Energy Materials," *Advanced Materials* 15/21 (2003): 1789-1803, 1795.
[178] Yonghong Kuang et al., "A Review of Renewable Energy Utilization in Islands," *Renewable and Sustainable Energy Reviews* 59 (2016): 504-513, 506.
[179] Daniel M. Kammen, "The Rise of Renewable Energy," 86.

Especially in rural areas, the use of thermal solar energy is more common. Photovoltaic (PV) solar energy is the generation of electricity from photons of sunlight by means of solar panels (solar cell panels). Photovoltaic energy systems can be used in many different places. It can be used on the roof and walls of houses, the surfaces of water sources, deserts, forest areas and even on the surfaces of electronic devices.[180]

Solar energy, which is most commonly used in electricity generation and water heating[181], has many advantages in terms of the environment. The most important advantage of solar energy is that no greenhouse gas is released while producing energy. From this point of view, solar energy has a significant advantage over fossil fuels in the fight against climate change. The second important advantage of solar energy in terms of the environment is that there is no hazardous waste during energy production. Therefore, solar energy has an important opportunity to prevent hazardous wastes caused by nuclear energy. However, as in other energy sources, environmental costs arise in solar energy. The use of the natural environment for solar panels negatively affects biodiversity.

3.2.2. Wind Power

Wind energy is the energy obtained by converting the kinetic energy of the wind into electricity.[182] Wind energy is the conversion of the motion energy of the air stream into electricity or energy by using wind turbines.[183] There are two basic types of wind turbines. These are vertical axis wind turbines and horizontal axis wind turbines.[184] The main source of wind energy is the sun. Because the pressure differences that arise due to the sun's heating of the atmosphere, seas and land parts at different

[180] Daniel M. Kammen, "The Rise of Renewable Energy," 86.

[181] Mesut Doğan, "Enerji Kullanımının Coğrafi Çevre Üzerindeki Etkileri," *Marmara Coğrafya Dergisi* 23 (2011): 36-52, 46.

[182] Stanley R. Bull, "Renewable Energy Today," 1217.

[183] Cumali İlkiliç, "Wind Energy and Assessment of Wind Energy Potential in Turkey," *Renewable and Sustainable Energy Reviews*, 16/2 (2012): 1165-1173, 1166.

[184] William P. Kuvlesky et al., "Wind Energy Development and Wildlife Conservation: Challenges and Opportunities," *The Journal of Wildlife Management*, 71/8 (2007): 2487-2498.

degrees create air movements, namely the wind.[185] Therefore, in order for the wind to exist, the sun must first exist.[186]

Although wind energy has made rapid progress in the last century, the history of the use of wind energy dates back to ancient times. Wind energy has been used for thousands of years to grind wheat in mills and to move sailing ships.[187] For example, wind energy was used to transport materials by boats in Egypt and Iraq 3500 years ago.[188] But the history of electricity generation from wind energy is relatively new. The first electricity generation from wind energy took place towards the end of the 19th century.[189] The first wind turbine was produced by Dane Poul LaCour in 1891.[190] Due to the potential of wind energy, its share in world energy production is expected to increase more.[191]

Wind energy is a clean energy source. No pollutant is released when wind energy is produced by wind turbines.[192] In addition, greenhouse gases are not released during energy production in wind energy.[193] For this reason, wind energy is considered one of the most effective energy sources in the fight against air pollution and climate change. However, wind energy also has some environmental risks. The first of these is the damage it causes to bird species.[194] Wind turbines located on the migration routes of birds pose a risk in protecting

---

[185] Nada Alrikabi, "Renewable Energy Types," 62.

[186] Ahmet Özen, Mahmut Ünsal Şaşmaz and Ercan Bahtiyar, "Türkiye'de Yeşil Ekonomi Açısından Yenilenebilir Bir Enerji Kaynağı: Rüzgâr Enerjisi," *Karamanoğlu Mehmetbey Üniversitesi Sosyal ve Ekonomik Araştırmalar Dergisi*, 17/28 2015: 85-93, 87.

[187] İbrahim Aydın, "Balıkesir'de Rüzgâr Enerjisi," *Doğu Coğrafya Dergisi*, 18/29 (2013): 29-50, 31.

[188] Selcuk Bilgen, Kamil Kaygusuz and Ahmet Sari, "Renewable Energy," 1125.

[189] Robert Gross, Matthew Leach and Ausilio Bauen, "Progress in Renewable Energy," 108.

[190] Mahmut Can Şenel and Erdem Koç, "Dünyada ve Türkiye'de Rüzgar Enerjisi Durumu-Genel Değerlendirme," *Mühendis ve Makina* 56/663 (2015): 46-56, 47.

[191] Matthew T. Huber and James McCarthy, "Beyond the Subterranean," 664.

[192] Agbetuyi A. Felix et al., "Wind Energy Potential in Nigeria," *International Electrical Engineering Journal*, 3/1 (2012): 595-601.

[193] R. Saidur et al., "Environmental Impact of Wind Energy," *Renewable and Sustainable Energy Reviews*, 15/5 (2011): 2423-2430.

[194] K. Shawn Smallwood, "Comparing Bird and Bat Fatality-Rate Estimates Among North American Wind-energy Projects," *Wildlife Society Bulletin*, 37/1 (2013): 19-33.

biodiversity.[195] At the same time, the sound generated during the rotation of wind turbines triggers noise pollution.[196]

3.2.3. Wave Energy

Wave energy is the energy generated by the conversion of waves in water resources into energy. Wave energy is obtained by converting the movement that occurs with the gravitational interaction of the earth and the moon into energy.[197] It is possible to accept it as a form of wind energy since wave energy emerges as a result of the movements created by the wind in water resources.[198] The potential of wave energy is quite high. Because the oceans cover 70% of the earth's surface, the potential of wave energy is greater than the potential of solar and wind energy.[199]

It is not a new process for people to benefit from wave energy. Wave energy has been used by people in different forms for hundreds of years. For example, it is known that waves were used to turn the wheels of mills in China in the 13th century.[200] The generation of energy by using advanced technological tools in wave energy is relatively new. The first patent application for a wave energy device was made in France in 1799.[201]

The use of wave energy as an energy source has significant advantages and disadvantages. It is possible to collect its advantages under two headings. First, there is no direct emission of greenhouse gases in

---

[195] Yakut Can Aşıkoğlu and Murat Tosunoğlu, "Rüzgâr Enerji Santral Sahalarındaki Kuşların Üreme Faaliyetlerinin Araştırılması," *Turkish Journal of Bioscience and Collections* 2/1 (2018): 12-26.

[196] Shifeng Wang and Sicong Wang, "Impacts of Wind Energy on Environment: A Review," *Renewable and Sustainable Energy Reviews* 49 (2015): 437-443.

[197] Robert Foster, Majid Ghassemi and Alma Cota, *Solar Energy*, 2.

[198] Markel Penalba, Giussepe Giorgi and John V. Ringwood, "Mathematical Modelling of Wave Energy Converters: A Review of Nonlinear Approaches," *Renewable and Sustainable Energy Reviews* 78 (2017): 1188-1207, 1189.

[199] Muhammet Kaan Yeşilyurt et al., "A Scrutiny Study on Wave Energy Potential and Policy in Turkey," *Periodicals of Engineering and Natural Sciences* 5/3 (2017): 286-297, 287.

[200] Iraide López et al., "Review of Wave Energy Technologies and the Necessary Power-equipment," *Renewable and Sustainable Energy Reviews* 27 (2013): 413-434, 414.

[201] Robert Gross, Matthew Leach and Ausilio Bauen, "Progress in Renewable Energy," 113.

wave energy.[202] It has a significant advantage in tackling climate change caused by fossil fuels. Secondly, it does not trigger air pollution because it does not cause the release of pollutants. Wave energy also has disadvantages. First, wave energy can be used in countries with large water resources such as the sea and ocean. Therefore, it may be insufficient to meet the world's energy needs. Secondly, since the intensity of the wave will differ from season to season, it is not possible to get the same efficiency from wind energy all year round.[203]

3.2.4. Hydroelectric Energy

Hydroelectric energy is the conversion of falling or flowing water into energy using the force of gravity.[204] Hydroelectric energy uses the power of moving water.[205] Converting flowing water to energy is not a new development. Water has been used as an energy source for 2000 years.[206] With the technological developments in the energy sector, the transformation of the movement of water into energy has been developed. Hydroelectric power plants emerged as a result of this development. Hydroelectric energy is an energy source that emerged with the technological development of water wheels in the Industrial Revolution.[207] It is possible to convert almost all of the kinetic energy generated by the water flowing through the tribunes in hydroelectric power plants into electricity.[208] From this point of view, it seem reasonable to say that hydroelectric power plants have high efficiency.

---

202 M. Fadaeenejad et al., "New Approaches in Harnessing Wave Energy: With Special Attention to Small Islands," *Renewable and Sustainable Energy Reviews*, 29 (2014): 345-354, 346-347.

203 Anto´nio F. de O. Falca˜o, "Wave Energy Utilization: A Review of the Technologies," *Renewable and Sustainable Energy Reviews* 14/3 (2010): 899-918, 901.

204 Askari Mohammad Bagher et al., "Hydroelectric Energy Advantages and Disadvantages," *American Journal of Energy Science*, 2/2 (2015): 17-20, 17.

205 Vedat Karadeniz, Erdal Akpınar and Adem Başıbüyük, "Nehir Tipi Hidroelektrik Santraller ve Çevresel Etkileri (Reşadiye Hidroelektrik Santralleri Örneği)," *Doğu Coğrafya Dergisi* 16/26 (2011): 95-114.

206 Asit K.Biswas, "Impacts of Hydroelectric Development on the Environment," *Energy Policy* 10/4 (1982): 349-354, 349.

207 Geoffrey P. Sims, "Hydroelectric Energy," *Energy Policy*, 19/8 (1991): 776-786, 776.

208 Edmund Henrich et al., "The Role of Biomass in a Future World Without Fossil Fuels," *Chemie Ingenieur Technik* 87/12 (2015): 1667-1685, 1671.

Two instruments are needed for the establishment of hydroelectric power plants. The first of these is the dam. The second is freshwater resources such as rivers, lakes, and streams.[209] The most common use of hydroelectric power plants is to build a dam over a water source to store water in a reservoir.[210] In this type of production, energy is produced continuously and regularly with the water accumulated in the reservoir formed by closing the front of the water with a dam.[211] Dam-type hydroelectric power plants are also called storage power plants.[212] Another type of hydroelectric power is the hydroelectric power plant, also known as the river type. Since there is no water storage in this type of power plant, it is also called a power plant without storage.[213] The amount of energy production in river-type hydroelectric power plants is at a lower level compared to the power plants with storage.

The use of hydroelectric energy is becoming increasingly common all over the world. Because hydroelectric energy has significant advantages compared to other energy sources. Hydroelectric energy has three important advantages. First, hydroelectric power is a renewable energy source.[214] It is possible to produce hydroelectric energy as long as there is kinetic energy created by the movement of water and water. Secondly, hydroelectric energy causes the emission of a low amount of greenhouse gases. Therefore, it is an important advantage in the fight against climate change. Third, the cost of power generation in

---

209 Mildred S. Dresselhaus and I. L. Thomas, "Alternative Energy Technologies," *Nature* 414/6861 (2001): 332-337, 332.

210 Askari Mohammad Bagher et al., "Hydroelectric Energy Advantages," 19.

211 Faruk Oral, Rasim Behçet and Kadir Aykut, "Hidroelektrik Santral Rezervuar Verilerinin Enerji Üretimi Amaçlı Değerlendirilmesi," *Bitlis Eren Üniversitesi Fen Bilimleri Dergisi* 6/2 (2017): 29-38, 31.

212 Uzun Osman, "Hidro Elektrik Santraller (HES) ve Çevresel Etki Değerlendirmesi (ÇED) Düzce Örneği," *Düzce Üniversitesi Orman Fakültesi Ormancılık Dergisi* 7/2 (2011): 1-13, 2.

213 İbrahim Yurtseven, "Nehir Tipi Hidroelektrik Santrallerinin Havzalar Üzerindeki Ekohidrolojik Etkileri," *Journal of the Faculty of Forestry Istanbul University* 61/1 (2011): 55-62.

214 Sadık Alashan, Zekai Sen and Z. Fuat Toprak, "Hydroelectric Energy Potential of Turkey: A Refined Calculation Method," *Arabian Journal for Science and Engineering* 41/4 (2016): 1511-1520.

hydroelectric power plants is low. However, the installation cost of hydroelectric power plants is high.[215]

On the other hand, as with other energy sources, hydroelectric energy has disadvantages, especially in terms of the environment. Hydroelectric power plants carry significant risks to the environment.[216] These risks can be examined under two headings. First, dams damage the river ecosystem. The destruction of limited freshwater resources due to hydroelectric power plants increases the risk of water scarcity[217]. Second, it causes lands to be flooded.[218] This environmental degradation caused by hydroelectric power plants harms biodiversity.[219]

3.2.5. Geothermal Energy

Geothermal energy can also be defined as thermal energy.[220] Geothermal energy is the energy found as heat in the interior of the earth.[221] Geothermal energy emerges by using the heat of underground water directly or converting it into energy. It is widely used in volcanic regions where underground heat is intense.[222] The geothermal water has been used for washing, cooking, treatment, and cleaning since ancient

---

215 Asit K.Biswas, "Impacts of Hydroelectric Development," 349.

216 Peter H. Gleick, "Environmental Consequences of Hydroelectric Development: The Role of Facility Size and Type," *Energy* 17/8 (1992): 735-747, 745.

217 The world has rich water reserves. 70% of the planet we live on consists of water. However, 97.5% of the world's water consists of salt water. The amount of fresh water that people can use for their basic needs constitutes approximately 2.5% of the total water resources. 87% of fresh water is found in glaciers, atmosphere, underground and soil, that is, it cannot be used. Therefore, only 0.3% of the available water resources in the world are usable (For detailed information, see the book entitled 100 Unknown Scientific Facts on Environmental Problems).

218 Po-An Su and d Bryan Karney, "Micro Hydroelectric Energy Recovery in Municipal Water Systems: A Case Study for Vancouver," *Urban Water Journal* 12/8 (2015): 678-690, 678.

219 Mehmet Berkün, Egemen Aras and Tuğçe Koç, "Barajların ve Hidroelektrik Santrallerin Nehir Ekolojisi Üzerinde Oluşturduğu Etkiler," *Türkiye Mühendislik Haberleri* 452/6 (2008): 41-48.

220 Omar Ellabban, Haitham Abu-Rub and Frede Blaabjerg, "Renewable Energy Resources," 751.

221 Enrico Barbier, "Geothermal Energy Technology and Current Status: An Overview," *Renewable and Sustainable Energy Reviews* 6/1-2 (2002): 3-65, 3.

222 Yonghong Kuang et al., "A Review of Renewable," 508.

times.[223] There is evidence that geothermal energy has been used in China for two thousand years.[224] But electricity generation from geothermal sources was first started by Prince Piero Ginori Conti in 1904 in Italy.[225] The commercial use of electricity produced from geothermal energy began in 1914.[226]

Geothermal energy is used in two basic forms. First, it is used to meet the heating needs of small settlements by means of a geothermal heat pump.[227] Geothermal heat pumps transfer the heat from geothermal resources to the residences and provide heating for the settlements, especially in the winter season. Hot water from geothermal sources is also used to meet the hot water needs of houses.[228] Second, geothermal resources are used in electricity generation. The USA, Philippines, Italy, Mexico, Indonesia, Japan, New Zealand, and Iceland are the countries that produce the most electricity from geothermal sources.[229]

Geothermal energy, which is considered a renewable and environmentally friendly energy source, has significant disadvantages. The most important of these is the limitation. It is not possible to find geothermal resources in every region of the world. Geothermal resources are only found in certain regions of the world.[230] Since countries such as the Philippines, El Salvador, New Zealand, Kenya, and Iceland have rich geothermal resources, they meet a significant part of their energy needs from geothermal energy.[231] However, since these resources are not

---

223 I. W. Johnston et al., "Emerging Geothermal Energy Technologies," *KSCE Journal of Civil Engineering* 15/4 (2011): 643-653, 643.
224 Ingvar B. Fridleifsson, "Geothermal Energy for the Benefit of the People," *Renewable and Sustainable Energy Reviews* 5/3 (2001): 299-312, 301.
225 Wim C. Turkenburg, "Renewable Energy Technologies," 255.
226 Ayhan Demirbas, "Global Renewable Energy Projections," *Energy Sources, Part B: Economics, Planning, and Policy* 4/2 (2009): 212-224, 216.
227 Emine Erol, "Türkiye'de Elektrik Enerjisinin Tarihi Gelişimi: 1902-2000" (Yayınlanmamış Doktora Tezi, İstanbul Üniversitesi Sosyal Bilimler Enstitüsü, 2007), 134.
228 Ayhan Demirbas, "Global Renewable Energy Projections," 216.
229 Ingvar B. Fridleifsson, "Geothermal Energy," 302.
230 Mildred S. Dresselhaus and I. L. Thomas, "Alternative Energy Technologies," 332.
231 "Geothermal Energy," last updated date: 15 November, 2021, https://www. irena.org/ geothermal

available in all regions of the world, it is not possible to meet the world's energy needs from geothermal energy.

3.2.6. Biomass Energy

Biomass is the general name of all living things in nature.[232] Biomass energy, on the other hand, is a renewable energy source obtained from waste.[233] Biomass energy is obtained by converting organic wastes into energy. But these wastes are non-fossil wastes.[234] The world has rich resources in terms of biomass energy. The annual bioenergy potential of the world is estimated to be about 2900 exajoules (EJ)[235]. In other words, it seems possible to meet the world's energy needs with only biomass energy.

There are three main sources of waste used in biomass energy.[236, 237] The first of these is animal waste. Feces of animals such as sheep, cattle, horses, and chickens are commonly used waste. The second main source is a vegetable waste. Oilseed plants such as soybean and sunflower, fiber plants such as flax and hemp, and all kinds of forest wastes from forests are used in biomass energy production. For this reason, forests are the most important source of biomass energy.[238] The third main source of biomass energy is an agricultural and urban waste. Organic wastes generated in homes, hotels, and factories are converted into biomass energy. Both dry and liquid wastes are used in biomass energy.[239] Due to its diversity,

---

[232] Ayhan Demirbaş, "Global Renewable Energy Resources," 781.

[233] Roger Sathre and Leif Gustavssonö "Time-dependent Climate Benefits of Using Forest Residues to Substitute Fossil Fuels," *Biomass and Bioenergy* 35/7 (2011): 2506-2516, 2506.

[234] Martin Kaltschmitt, Wolfgang Streicher and Andreas Wiese (eds.), *Renewable Energy*, 4.

[235] Joule (J) is a unit of heat/energy measurement. EJ is equivalent to one quintillion (1018) joules. The amount of energy consumed in the world in 2020 is around 556 exajoules.

[236] Volkan Ş. Ediger and Elçin Kentel, "Renewable Energy Potential as an Alternative to Fossil Fuels in Turkey," *Energy Conversion and Management* 40/7 (1999): 743-755, 746.

[237] Nada Alrikabi, "Renewable Energy Types," 62.

[238] Juan JoséVidal-Amaro et al., "Optimal Energy Mix for Transitioning from Fossil Fuels to Renewable Energy Sources–The Case of the Mexican Electricity System," *Applied Energy* 150 (2015): 80-96, 86.

[239] Jamie Speirs, Christophe McGlade and Raphael Slade, "Uncertainty in the Availability," 661.

sources of biomass energy are more common than fossil fuels.[240]

Biomass energy is widely used in three forms.[241] First, heat is obtained from biomass energy. Biomass energy, which is widely used by the poor because of its easy and cheap access, is used for cooking and heating.[242, 243] The energy obtained by the direct use of biomass is also called traditional biomass energy.[244] Second, biomass energy is used to generate electricity. Third, biomass energy is converted into liquid fuels. These liquid fuels are used in vehicles.[245] The conversion of biomass as a result of various processes and obtaining energy is called modern biomass energy.[246]

The use of biomass energy provides significant advantages in protecting the physical and biological elements of the environment.[247] These advantages can be grouped under two headings. First, biomass energy is considered a carbon-neutral energy source.[248] Therefore, it can contribute to reducing the emission of greenhouse gases caused by fossil fuels. Secondly, biomass energy contributes to the recycling of wastes as energy.[249] Thanks to biomass energy, there is no need to transform the natural and artificial environment into landfills for waste disposal.

4. ENERGY HISTORY

Energy resources have been widely used all over the world after

---

[240] de Paula Protásio et al., "Brazilian Lignocellulosic Wastes for Bioenergy Production: Characterization and Comparison with Fossil Fuels," *BioResources* 8/1 (2013): 1166-1185, 1116.

[241] Omar Ellabban, Haitham Abu-Rub and Frede Blaabjerg, "Renewable Energy Resources," 750.

[242] Christian Azar et al., "Carbon Capture and Storage from Fossil Fuels and Biomass–costs and Potential Role in Stabilizing the Atmosphere," *Climatic Change* 74/1 (2006): 47-79, 54.

[243] Virginia H. Dale, Rebecca A. Efroymson and Keith L. Kline, "The Land Use Climate Change Energy Nexus," *Landscape Ecology* 26/6 (2011): 755-773, 756.

[244] Yonghong Kuang et al., "A Review of Renewable," 508.

[245] Nada Alrikabi, "Renewable Energy Types," 61.

[246] Yonghong Kuang et al., "A Review of Renewable," 508.

[247] de Paula Protásio et al., "Brazilian Lignocellulosic," 1176.

[248] Mohammad S. Masnadi et al., "From Fossil Fuels," 197.

[249] Omar Ellabban, Haitham Abu-Rub and Frede Blaabjerg, "Renewable Energy Resources," 751.

the Industrial Revolution. Because with the Industrial Revolution, human needs began to change.[250] However, the use of energy in human history dates back to before the Industrial Revolution. The first development in which humans benefited from energy was the invention of fire. With the invention of fire, energy was obtained by burning wood and other biomass energy sources.[251] The heat energy obtained in this period was used for different purposes. The first of these is to be used in cooking. The second is for heating purposes. The third was used to protect themselves against wild animals. Due to the widespread use of wood, it is estimated that a significant amount of deforestation occurred in Europe in the period from 2000 BC to 1800 AD.[252]

Another of the oldest energy sources used for human needs in primitive times is renewable energy sources. The oldest renewable energy source used by humans is wind energy which has been widely used in two areas. First, the sailing ships used for transportation were operated from wind energy. Second, the wind was converted into energy by means of ground mills. The energy produced from water through waterwheels has been used since ancient times.[253]

In the First Age and the Middle Ages, the use of fossil fuels began to become widespread. Coal was widely used in this period due to its easy accessibility. Since there is no advanced technology, the usage areas of coal have been limited. Coal was most commonly used to supply hot water for heating and cleaning needs in homes. Coal has also been used for cooking. During this period, coal use is quite common, especially in European countries. England, located in geography rich in coal reserves, is one of the countries with the highest coal use. Due to the widespread use of coal in these periods, deadly environmental problems arose in England. In order to prevent air pollution caused by the use of coal, King

---

[250] Emrah Atar, "Managing Air Pollution in International Environmental Policies: Covid-19 as a Case to Demonstrate the Impact of Human on Environment," *Journal of Academic Value Studies* 7/2 (2021): 63-72.

[251] David Bodansky, Nuclear Energy: Principles, 1.

[252] Jed O. Kaplanab, Kristen M. Krumhardta and Niklaus Zimmermann, "The Prehistoric and Preindustrial Deforestation of Europe," *Quaternary Science Reviews* 28/27-28 (2009): 3016-3034.

[253] Aviel Verbruggen, "Renewable and Nuclear Power: A Common Future?," *Energy Policy* 36/11 (2008): 4036-4047, 4038.

Edward of England made legal regulations prohibiting the use of coal in the 14th century as stated before.[254]

Until the 18th century, there was no significant change in the world's energy outlook. However, with the Industrial Revolution, world energy consumption has entered into a rapid increase.[255] The use of energy resources in production, transportation, and residences has triggered a rapid increase in energy production and consumption. After the Industrial Revolution, the most important energy source in the world is coal. In the first half of the 19th century, coal consumption reached the highest level in human history. Until 1947, approximately 81 billion metric tons of coal were consumed in human history. 40 billion metric tons of this were used between 1920-and 1947, and 62 million metric tons of it were used between 1900-and 1927.[256]

After the Industrial Revolution, other types of fossil fuels began to be used. Besides coal, oil and natural also started to be used in industries, transportation, and residences. Conventional petroleum was first discovered in 1859.[257] The first natural gas well in the world was drilled in the USA in 1825.[258] The fuels produced by the processing of oil and natural gas have been widely used in residences, factories, and vehicles. Especially in the 20th century, the use of oil and natural gas began to increase rapidly.

Since the 1950s, the environmental effects of fossil fuels have been discussed more in the world. Because of the environmental pollution caused by fossil fuels, public health has been adversely affected and mass deaths have occurred. One of these mass deaths occurred in London, the center of Europe. More than 4000 people died in London[259], the capital of England, in 1952 due to air pollution caused by excessive use of coal. Due to the environmental problems reaching dangerous levels, the idea of

---

254 Philip K. Hopke, "Contemporary Threats and Air Pollution," *Atmospheric Environment* 43/1 (2009): 87-93, 87.

255 Anne E. Maczulak, *Renewable Energy: Sources and Methods* (New York: Infobase Publishing, 2010), 2.

256 M. King Hubbert, "Energy from Fossil Fuels," *Science* 109/2823 (1949): 103-109, 103.

257 Naeem Abas, A. Raza Kalair and Nasrullah Khan, "Review of Fossil Fuels and Future Energy Technologies," *Futures* 69 (2015): 31-49, 5.

258 Kenneth J. Chew, "The Future of Oil," 3.

259 Murat Çetin and Oğuz Kürşat Demirci, "Erzincan'da Doğal Gaz," 11.

reducing the use of fossil fuels and using alternative energy sources has started to receive more support on a global scale.

Nuclear energy is the most important energy source that emerged as an alternative to fossil fuels in the 1950s. Among the energy sources, the most recently discovered or the newest energy source is nuclear energy. Nuclear energy, which has been used since the 1950s, is a new type of energy source compared to fossil fuels and renewable energy sources that have been used for thousands of years. Nuclear energy has been used in energy production since the middle of the 20th century.[260] During the Cold War years, the competition between the Soviet Union and the USA in every field was also seen in nuclear energy technology. As a result of this competition, nuclear energy technology developed rapidly and started to be used in other countries. Investment in nuclear energy gained momentum, especially with the fossil fuel crisis that started to be seen in the 1970s.

Since energy is limited in nature, in the second half of the 20th century, the world began to experience significant energy crises. Problems in supplying energy from countries with fossil fuels have led to energy crises. These energy crises are as follows; (1) The 1973 Oil Crisis, (2) The 1979 Energy Crisis, and (3) The 1990 Oil-price Hike.[261] With the oil-based energy crises, the world's energy outlook has begun to change. Difficulty in access to oil and rapid rise in oil prices have increased the tendency towards alternative energy sources. One of these alternative energy sources is nuclear energy. As a result of the oil crisis that emerged in the 1970s, the share of nuclear energy in electricity production began to increase.[262]

With the energy crisis in the 1970s and the development of social awareness about environmental pollution caused by fossil fuels, renewable energy has also come to the fore.[263] In addition, the environmental disaster that occurred as a result of the accident at the nuclear power plant in

---

[260] David Bodansky, Nuclear Energy: Principles, 1.

[261] V.Manieniyan, M. Thambidurai and R. Selvakumar, "Study on Energy Crisis and the Future of Fossil Fuels," *Proceedings of SHEE* 10 (2009): 2234-3689.

[262] Ferenc L. Toth and Hans-Holger Rogner, "Oil and Nuclear Power: Past, Present, and Future," *Energy Economics* 28/1 (2006): 1-25, 5.

[263] Atte Harjanne and Janne M. Korhonen, "Abandoning the Concept," 331.

Chernobyl[264] has increased the interest in renewable energy sources, which are accepted as an environmentally friendly energy source. Renewable energy sources have been used more and more all over the world since the end of the 20th century. As a result of the rapid technological progress in the field of renewable energy, the share of renewable energy sources such as solar energy, wind energy, wave energy, thermal energy, and hydroelectric energy in energy production has tended to increase rapidly.

When it comes to today, it is possible to say that the diversity of energy sources has increased. The dependency on a single energy source, which was seen until the 1950s, is not in question today. The world continues to meet a significant portion of its energy needs from fossil fuels. However, the share of renewable energy sources and nuclear energy is in a steady increasing trend. While the use of fossil fuels is decreasing, the use of renewable energy sources is increasing rapidly. In addition, it is estimated that the share of renewable energy sources will increase much more in the coming years.

5. WORLD ENERGY OUTLOOK

Three main energy sources are used in the production of the energy needed by the world.[265] The first and most common of these are fossil fuels. The world is dependent on fossil fuels for energy production.[266] Because fossil fuels meet approximately 83% of the world's energy needs. Renewable energy is the second most important energy source after fossil fuels in energy production. Approximately 13% of the energy consumed in the world is obtained from renewable energy sources such as solar energy, wave energy, wind energy, and hydroelectric energy. Nuclear energy has the lowest share in energy production in the world. About 4% of the world's energy needs are met from nuclear energy.

---

[264] Ilya A. Likhtarev et al., "Chernobyl Accident: Retrospective and Prospective Estimates of External Dose of the Population of Ukraine," *Health Physics* 82/3 (2002): 290-303.
[265] Ali Mirchi et al., "World Energy Balance Outlook and OPEC Production Capacity: Implications for Global Oil Security," *Energies* 5/8 (2012): 2626-2651, 2628.
[266] Volkan Ş. Ediger, "An Integrated Review," 3.

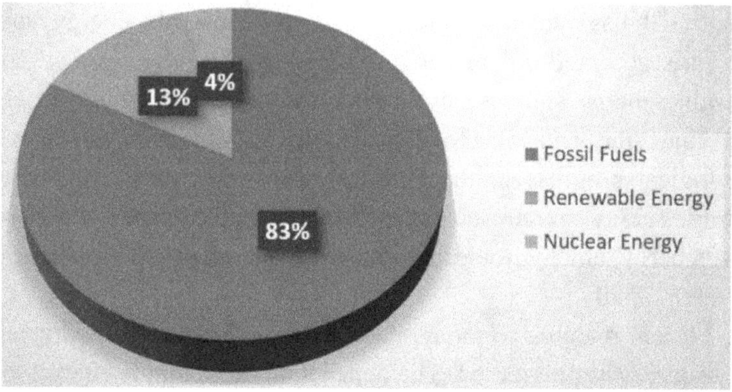

Table 4: Ratio of Energy Resources in World Energy Production in 2020 [267]

In the last 300 years, the most widely used energy source in the world is coal. Coal meets 36.7% of the world's energy needs. As seen in Table 4, 97 terawatt-hours of energy were produced from coal in 1800. This rate increased significantly in 1900 and reached 5728 terawatt-hours. The gradual increase continued until 1950. In 1950, the energy produced from coal in the world was 12603 terawatt-hours. In the period from 1950 to the present, there has been a sharp increase in the amount of energy produced from coal. In 2019, 50,000 terawatt-hours of energy were produced from coal.

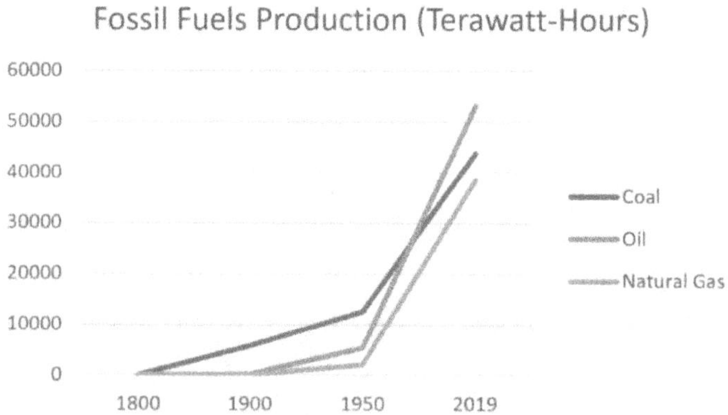

Table 5: Fossil Fuel Production in the World Between 1800-2019[268]

[267] "Statistical Review of World Energy 2021," 12.

[268] "Fossil Fuels," last updated date: 30 March, 2022, https://ourworldindata.org/fossil-fuels

Oil is the most widely used energy source after coal among fossil fuels. Energy production from oil is relatively new compared to coal. Towards the end of the 1800s, the oil began to be used in energy production. In 1900, 181 terawatt-hours of energy were produced from oil. This rate increased significantly by 1950. In 1950, 5444 terawatt-hours of energy were produced from oil. Since the 1950s, the rapid increase in the use of oil has continued. In 2019, 53,181 terawatt-hours of energy were produced from oil.

Natural gas is the third most common energy source among fossil fuels.[269] Natural gas, like oil, started to be used in energy production towards the end of the 1800s. In 1900, 64 terawatt-hours of energy were produced from natural gas. At the beginning of the 20th century, the share of natural gas in energy production was quite low. However, there was a rapid increase in energy production from natural gas until 1950. In 1950, 2092 terawatt-hours of energy were produced from natural gas. This increase continued until 2019, gaining momentum. In 2019, 39,292 terawatt-hours of energy were produced from natural gas.

While the amount of energy produced from fossil fuels increases[270], the share of fossil fuels in total energy production tends to decrease. One of the main reasons for this is the increasing population. The world population is increasing.[271] Especially since the 1950s, the world population has started to increase uncontrollably. As seen in Table 3, the world population, which was around 1.6 billion in 1900, reached 2.54 billion in 1950. The world population, which continues to increase, became 7.71 billion in 2019. It is estimated that the world population will exceed 10 billion by 2050.[272]

---

[269] Tomas Kåberger, "Progress of Renewable Electricity," 50.

[270] Shahriar Shafiee and Erkan Topal, "An Econometrics View," 777.

[271] Emrah Atar, "Birleşmiş Milletler Dünya Gıda Programı'nın Evrensel Gıda Sistemi ve Dağıtımındaki Rolünün Analizi," Disiplinlerarası Gıda Çalışmaları Dergisi 1/1 (2021): 45-61.

[272] Selcuk Bilgen, Kamil Kaygusuz and Ahmet Sari, "Renewable Energy," 1119.

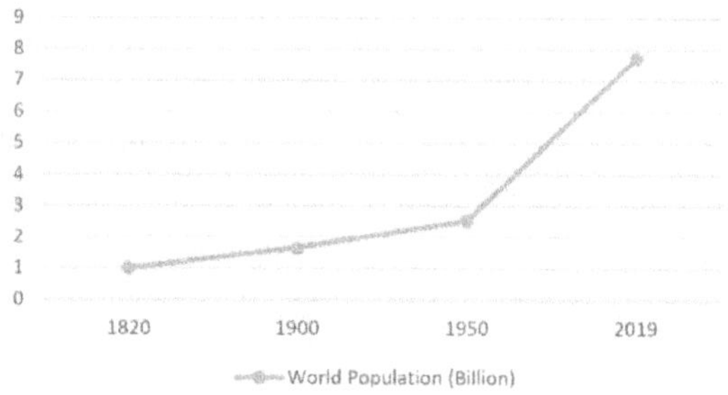

Table 6: World Population Rates for 1820-2019 [273]

Due to the increase in the human population, the need for energy also increases in parallel.[274] As more people are born, more energy is consumed. Because energy is consumed continuously in many phases of our daily life from the first year we are born. The increase in the need for energy along with the increase in the population makes it necessary to use more energy sources. However, fossil fuels are non-renewable energy sources that are found in nature in limited quantities.[275] As the population increases, the amount of fossil fuels decreases.[276] For this reason, it is insufficient to meet the energy needs of the rapidly increasing world population. This situation necessitates the use of alternative energy sources.

Another factor that necessitates the use of alternative energy sources[277] to fossil fuels is the environmental pollution caused by fossil fuels.[278] It is the most polluting type of energy among the energy sources. As explained in the previous sections, greenhouse gases are released into the atmosphere with the burning of fossil fuels. These pollutants, which cause air pollution, trigger climate change, which is one of the most important environmental problems of the period we live in.[279] For these

---

[273] "World Population Growth," last updated date: 31 March, 2022 https://ourworldindata .org/world-population-growth

[274] Richard York, "Do Alternative Energy Sources Displace Fossil Fuels?," Nature Climate Change 2/6 (2012): 441-443, 443.

[275] V.Manieniyan, M. Thambidurai and R. Selvakumar, "Study on Energy."

[276] Ümit Ağbulut and Suat Sarıdemir, "A General View," 1569.

[277] Alicia Valero and Antonio Valero, "What are the Clean," 130.

[278] Cees Withagen, "Pollution and Exhaustibility," 235.

[279] Govinda R. Timilsina, Stefan Csordás and Simon Mevel, "When Does a Carbon Tax on Fossil Fuels Stimulate Biofuels?," *Ecological Economics* 70/12 (2011): 2400-2415.

reasons, priority has been given to alternative energy sources to fossil fuels since the 1950s.

There are two main types of energy sources emerging as alternatives to fossil fuels.[280] One of these energy sources is nuclear energy. The first nuclear power plant producing nuclear power was established in Obninsk, Soviet Union, in 1954.[281] It started in the 1980s with the spread of nuclear energy. As can be seen in Table 7, energy production from nuclear energy has approximately doubled between 1980 and 2006. However, it is seen that the demand for nuclear energy is in a decreasing trend between the years 2006-2012. The reason for this is the decrease in nuclear energy production in countries such as Germany, the USA, and France after the Fukushima Nuclear Power Plant accident that occurred in the Japanese state of Fukushima in 2011.[282] It is seen that there is an increase in nuclear energy between 2012 and 2020. Nuclear energy currently meets 4% of the world's energy needs.

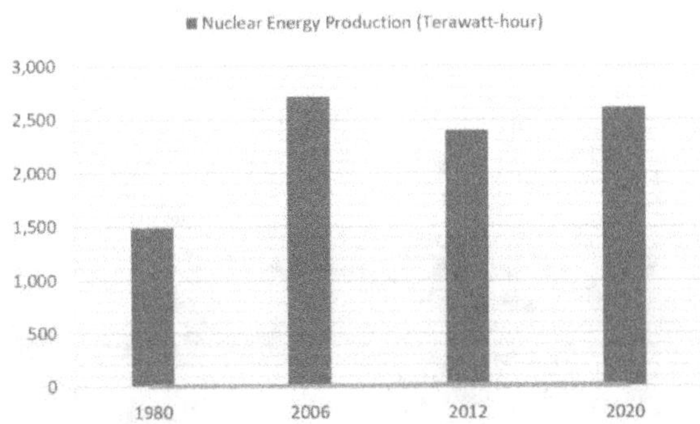

Table 7: Nuclear Energy Production in the World Between 1980-2020[283]

---

[280] Thomas Covert, Michael Greenstone and Christopher R. Knittel, "*Will We Ever Stop Using Fossil Fuels?*," *Journal of Economic Perspectives* 30/1 (2016): 117-38, 120.

[281] V. I. Rachkov et al., "From the First Nuclear Power Plant to Fourth-generation Nuclear Power Installations [on the 60th Anniversary of the World's First Nuclear Power Plant]," *Thermal Engineering* 61/5 (2014): 327-336, 327.

[282] Claudia Furlan, Mariangela Guidolin and Renato Guseo, "Has the Fukushima Accident Influenced Short-term Consumption in the Evolution of Nuclear Energy? An Analysis of the World and Seven Leading Countries," *Technological Forecasting and Social Change* 107 (2016): 37-49, 48.

[283] "Nuclear Energy," last updated date: 5 April, 2022, https://ourworldindata.org/nuclear-energy

With the Chernobyl Nuclear Power Plant accident that occurred in Ukraine in 1986, alternative energy sources to nuclear energy started to come to the fore. Since the second half of the 20th century, due to the environmental effects of fossil fuels and nuclear energy, renewable energy sources have started to take a larger share in world energy production. As seen in Table 8, energy production from renewable energy sources has been increasing gradually since 1965. In 1965, 941 terawatt-hours of energy were produced from renewable energy sources. This ratio reached 1781 terawatt-hours of energy in 1980, 2280 terawatt-hours of energy in 1990, and 2870 terawatt-hours of energy in 2000, respectively. From 2000 to 2010, the energy produced from renewable energy sources has nearly doubled, that is, to 4197 terawatt-hours. This rapid increase continues until today. In 2020, the energy produced from renewable energy sources reached 7444 terawatt-hours of energy.

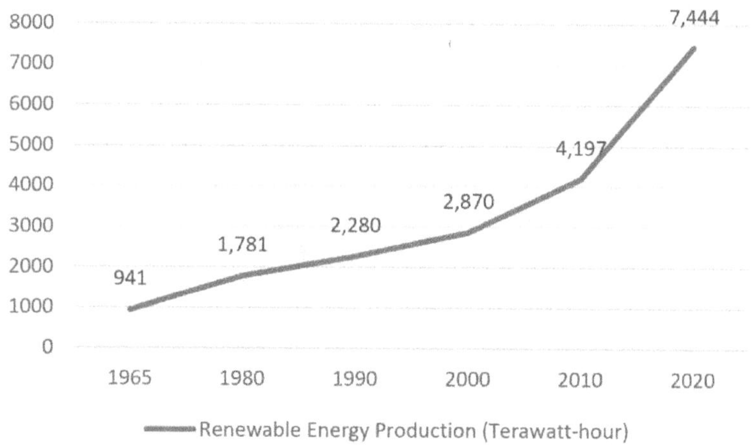

Table 8: Renewable Energy Production in the World Between 1965-2020[284]

The share of fossil fuels, renewable energy sources, and nuclear energy in energy production has been in a gradual tendency for the last 50 years. However, there was no significant change in dependence on fossil fuels from the Industrial Revolution period to the second half of the 20th century. From the Industrial Revolution, which started with the invention

284 "Renewable Energy," last updated date: 7 April, 2022, https://ourworldindata.org/renewable-energy

of steam engines using fossil fuels by the British engineer Thomas Savery in 1698, until the 1980s, almost all of the energy production was met from fossil fuels. With the rapid change in technology in the last 50 years, the share of nuclear energy and renewable energy sources in energy production has started to increase against fossil fuels. As seen in Table 9, the energy produced from fossil fuels is in an increasing trend between 1985 and 2020. A similar increase is also seen in renewable energy sources. There is a partial increase in nuclear energy. In the last 50 years, the share of fossil fuels in total energy production has tended to decrease due to other energy sources.

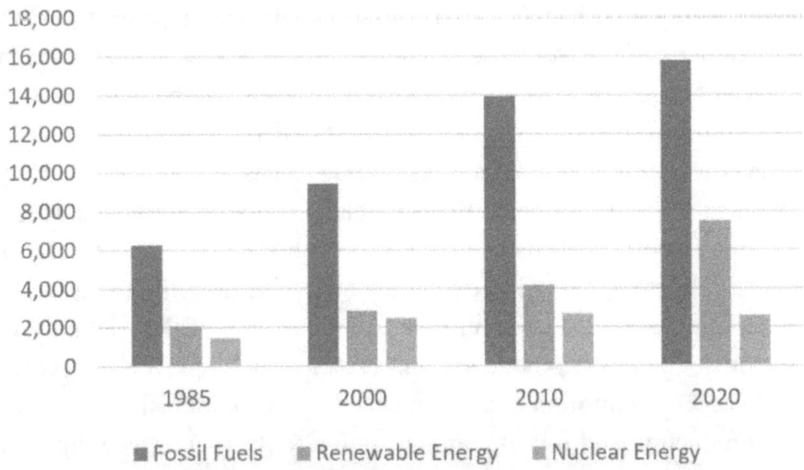

Table 9: Electricity Production Amount by Type of Energy Resources Between 1985-2020[285]

In summary, fossil fuels have the most important share in the world's energy outlook. The dependency on fossil fuels that started with the Industrial Revolution still continues. However, despite the increase in the energy produced from fossil fuels, its share in world energy production is gradually decreasing. Because countries have been turning to alternative energy sources, especially thanks to technological developments since the 1980s. One of them is renewable energy sources. The share of renewable

[285] "Electricity Production by Source," last updated date: 9 April, 2022, World, https://ourworldindata.org/grapher/electricity-prod-source-stacked

energy sources in world energy production is increasing rapidly.[286] Nuclear energy, which has the lowest share in energy production, has not had a remarkable increase in energy production in recent years, unlike renewable energy sources. This information shows that although the use of renewable energy sources increases in the medium and long term, the superiority of fossil fuels among energy sources in the short and medium-term will continue in the long term.[287, 288]

6. CONCLUSION

In this section, the scope of energy, the importance of energy, types of energy, the change of energy use in the historical process, and the energy outlook of the world are examined. First of all, what energy means and in which areas it is used are discussed. Energy can be defined as the capacity or power to do a job. The source of energy can be based on the muscle power of humans, natural resources such as the sun and wind, and the muscle power of animals. Energy obtained from different sources is one of the important needs of modern society. Energy is widely used in the following areas; (1) in increasing the quality of life of individuals, (2) in economic activities that contribute to the increase of the welfare level of societies, (3) in the production and use of technological products that provide access to information, (4) in ensuring technological progress, (5) in the production and use of communication tools, (6) in meeting basic needs such as cooking and heating, (7) in the production, processing, and transportation of consumer goods, and (8) in the production and use of individual and public vehicles.

Secondly, what are the energy sources, their reserves, and important risks and opportunities in terms of environment and public health are examined. Energy is obtained from different sources. These resources are categorized in three groups. The first of these is fossil fuels.

---

[286] Paolo Zeppini and Jeroen CJM Van Den Bergh, "Global Competition Dynamics of Fossil Fuels and Renewable Energy under Climate Policies and Peak Oil: A Behavioural Model," *Energy Policy* 136 (2020): 1-24, 1.

[287] Leif Gustavsson et al., "Climate Change Effects," 614.

[288] Nazim Muradov, "Hydrogen via Methane Decomposition: An Application for Decarbonization of Fossil Fuels," *International Journal of Hydrogen Energy* 26/11 (2001): 1165-1175.

Fossil fuels, consisting of coal, oil, and natural gas, are composed of organic compounds found in living things. Fossil fuels are a limited source of energy in nature and cannot be renewed in case of depletion. The second main energy source is renewable energy which constantly renews itself. Renewable energy emerges with the natural cycles of the sun and the universe. The most common types of them are solar energy, wind energy, wave energy, geothermal energy, biomass energy, and hydroelectric energy. The third main energy source is nuclear energy. The main fuel of nuclear energy, which has been used since the 1950s, is uranium. Nuclear energy is the energy produced by the fission of uranium. Nuclear energy is a non-renewable energy source due to the limited availability of uranium in nature.

Thirdly, how the use of energy resources has changed in the historical process has been examined in general terms. The use of energy resources has changed depending on the technological developments experienced in the historical process. People used renewable energy sources such as biomass energy, wind energy, and solar energy in the primitive period. With the discovery of fossil fuels, the use of coal has begun to replace biomass energy. In the Antiquity and the Middle Ages, the use of coal for heating and cooking spread rapidly in European countries. With the Industrial Revolution, the world's energy outlook has changed. Energy production and consumption have increased rapidly due to reasons such as the start of mechanization in the factories, the spread of mass production, and services due to the rapid increase in the world population. With the technological developments after the Industrial Revolution and the use of oil and natural gas, fossil fuels have become the main energy source in the world. Since the 1950s, alternative energy sources have started to be used due to the environmental effects of fossil fuels and the risk of depletion. The use of nuclear energy and renewable energy has started to increase since the second half of the 20th century.

Fourth and lastly, the energy view of the world has been examined. The world is dependent on fossil fuels for energy. Since the Industrial Revolution, the main energy source used in the world is fossil fuels. Approximately 83% of the world's energy needs are met from fossil fuels. The second most common energy source used in the world is renewable energy. Since the 1970s, the use of renewable energy sources has started to increase gradually. 13% of the world's energy needs are met by renewable energy. The third main energy source used in the world is nuclear energy. Nuclear energy, which has been used since the 1950s, meets 4% of the world's energy needs. While the use of fossil fuels is decreasing in the world, the use of renewable and nuclear energy tends to increase.

CHAPTER III:
NUCLEAR ENERGY

1. INTRODUCTION

Nuclear energy is a newly emerged energy source compared to fossil fuels and renewable energy. Fossil fuels and renewable energy sources have been used since ancient times in human history. With the economic and social transformation experienced after the Industrial Revolution, the rapid progress in the technological field has also changed the scope of energy resources. With the use of rapidly developing technology in the field of energy, alternative energy sources have emerged. One of these alternative energy sources is nuclear energy obtained from uranium.

Nuclear energy is one of the most discussed energy sources in the world. Because nuclear energy has important risks and opportunities in terms of the environment and public health. At the same time, its economic advantages and disadvantages are also frequently discussed. However, in order to make these discussions on a healthy ground, it is necessary to know what nuclear energy is, its scope, and how it has developed in the historical process.

The main purpose of this chapter is to provide descriptive information about the scope of nuclear energy. This chapter consists of three main parts. In the first part, what nuclear energy is, the main fuel of nuclear energy, and the working principle of nuclear energy are examined. In the second part, the history of nuclear energy is discussed. How nuclear energy emerged and developed is introduced. In the third and last part, the use of nuclear energy in the world and the nuclear energy policies of the countries are examined.

2. NUCLEAR ENERGY

The term nuclear comes from the English word "nucleus."[289] The word "nucleus," used in physics, means "the central part of an atom, usually composed of protons and neutrons," according to the Cambridge Dictionary.[290] The central part of the atom is also called the atomic nucleus. From this point of view, nuclear energy means atomic energy or atomic nucleus energy. In other words, nuclear energy is the energy source obtained from the atom. To put it more concretely, nuclear energy is the conversion of heat from the division of uranium by colliding with neutrons into energy.

The main fuel used in nuclear power plants is uranium. Uranium, which was discovered by the German chemist Kloproth in 1789[291], can only be extracted from solid ores.[292] Uranium, which is a radioactive element[293], is intensely found in saltwater sources. It is known that there are 4.5 billion tons of uranium in sea and ocean water.[294] Natural uranium consists of two isotopes. The first of these is Uranium-238 (U^{238}), which is the most abundant in nature. Since it is an indivisible isotope, it cannot be used in nuclear reactors.[295] The second is Uranium-235 (U^{235}). Today, Uranium-235 and enriched uranium are most commonly used in nuclear power plants.[296] Until 2010, 2.5 million tons of uranium were mined. 2

---

[289] Kadir Temurçin and Alpaslan Aliağaoğlu, "Nükleer Enerji ve Tartışmalar Işığında Türkiye'de Nükleer Enerji Gerçeği" *Coğrafi Bilimler Dergisi* 1/2 (2003): 25-39, 26.

[290] "Meaning of Nucleus in English," last updated date: 9 February, 2022, https://dictionary.cambridge.org/dictionary/english/nucleus

[291] Kadir Temurçin and Alpaslan Aliağaoğlu, "Nükleer Enerji ve Tartışmalar," 29.

[292] Claude Degueldre, "Uranium as a Renewable for Nuclear Energy," *Progress in Nuclear Energy* 94 (2017): 174-186, 174.

[293] Peter T. Todorov and Elica Nikolova Ilieva, Contamination with Uranium from Natural and Antropological Sources, *Romanian Journal of Physics* 51/1-2 (2006): 27-34, 27.

[294] Sophie Gabriel et al., "A Critical Assessment of Global Uranium Resources, Including Uranium in Phosphate Rocks, and the Possible Impact of Uranium Shortages on Nuclear Power Fleets," *Annals of Nuclear Energy* 58 (2013): 213-220, 214.

[295] QiangYan et al., "Nuclear Power Development in China and Uranium Demand Forecast: Based on Analysis of Global Current Situation," *Progress in Nuclear Energy* 53/6 (2011): 742-747.

[296] Bernd Lehmann, "Uranium Ore Deposits," *Reviews in Economic Geology* 2 (2008): 16-26, 17.

million tons of this was used in electricity generation and the remaining 500 thousand tons were used for military purposes.[297]

Nuclear energy emerges as a result of the nuclear fuel cycle. The nuclear fuel cycle is a long and complex process.[298] This cycle consists of three stages. The first stage is the "initial stage." The initial stage is the process of making the nuclear fuel ready. Some of the activities done in the beginning phase are as follows; mining, conversion, enrichment, and fuel production. The second stage covers the process of generating energy from nuclear fuel. There are two basic methods used in the fuel cycle.[299] The first of these methods is the once-through cycle. In this method, the uranium used in nuclear power plants is stored as waste in storage facilities. The second method is the closed fuel cycle. In this method, the plutonium formed during energy production is separated from the spent fuel and reused as fuel. The third step in the fuel cycle is the back end. The last stage covers the activities carried out after the nuclear fuel is taken from the nuclear power plant. Some of them are as follows; nuclear waste storage, nuclear waste reprocessing, and nuclear waste disposal.[300]

The installation of nuclear power plants, where uranium is converted into energy, goes through three basic stages.[301] The first of these is the construction of nuclear power plants. Choosing the right location before the construction of nuclear power plants is of strategic importance. Site selection for nuclear power plants takes place by adhering to certain criteria. The first is that it is close to water sources. Because a large amount of water is used in the production of nuclear energy. Nuclear power plants should be built close to water resources in order to meet the water demand regularly.

---

[297] Michael Dittmar, "The End of Cheap Uranium," *Science of the Total Environment* 461: 792-798, 793.

[298] Benjamin K. Sovacool, "Valuing the Greenhouse Gas Emissions from Nuclear Power: A Critical Survey," *Energy Policy*, 36/8 (2008): 2950-2963, 2951.

[299] Chaoran Zhou et al., "Economic Analysis of Two Nuclear Fuel Cycle Options," *Annals of Nuclear Energy* 71 (2014): 230-236, 230.

[300] Xiaoqi Sun, Huimin Luo and Sheng Dai, "Ionic Liquids-based Extraction: A Promising Strategy for the Advanced Nuclear Fuel Cycle," *Chemical Reviews* 112/4 (2012): 2100-2128, 2100.

[301] Benjamin K. Sovacool, "Valuing the Greenhouse Gas," 2951.

The second criterion to be considered in the selection of a site for the installation of nuclear power plants is to avoid areas where the risk of natural disasters is high. Because in regions where the risk of natural disasters is high, nuclear power plants may face high-security risks. For example, the construction of a nuclear power plant in a region with an earthquake fault line may pose a security problem. The security risk caused by natural disasters has started to come to the fore with the nuclear accident at the Fukushima Nuclear Power Plant in Japan in 2011. The tsunami that emerged after the earthquake with a magnitude of 9 damaged the nuclear power plants and triggered the nuclear accident.[302] For this reason, before the construction of nuclear power plants, the risks posed by natural disasters should be calculated correctly when choosing a location.

The second stage is the operation of nuclear power plants, that is, the production stage of nuclear energy.[303] How nuclear power plants are operated may differ depending on the type of nuclear power plant. Because there are different types of nuclear energy reactors, and their working principles may also differ. Since nuclear energy technology is under continuous development, different nuclear energy reactors have been used since the 1950s. Five different types of nuclear power reactors are most commonly used. These are as follows:[304]

- **The light water reactor:** The most widely used nuclear power reactor is the light water reactor. Until 1987, 84% of the reactors in the world consisted of light water reactors.[305] Nuclear power plants operating with light water reactors are cooled by water with low deuterium[306] value. Water is also used to slow down the neutrons necessary for the fission chain

[302] Carola Hommerich, "Trust and Subjective Well-being After the Great East Japan Earthquake, Tsunami and Nuclear Meltdown: Preliminary Results," *International Journal of Japanese Sociology* 21/1 (2012): 46-64, 46.

[303] Benjamin K. Sovacool, "Valuing the Greenhouse Gas," 2951.

[304] Kathleen Vaillancourt et al., "The Role of Nuclear Energy in Long-term Climate Scenarios: An Analysis with the World-TIMES model," *Energy Policy* 36/7 (2008): 2296-2307, 2298-2299.

[305] B. Pershagen and M. Bowen, *Light Water Reactor Safety* (Oxford: Pergamon Press, 2013), 1.

[306] Deuterium, which is heavy hydrogen, is one of the isotopes of hydrogen. Water with depleted or low levels of deuterium is called light water.

reaction to occur during energy production. The water used in these reactors comes out as high-pressure water. Light water reactors are divided into two groups. The first of these is pressurized water reactors. The second is the boiling water reactors.[307] In light water reactors, fuel cycles occur in two ways. In the first, enriched uranium fuels the nuclear reactor. It is then processed in a single pass cycle for long-term disposal. In the second, there is the closed-loop, where nuclear fuels using uranium and plutonium are recovered and used.[308] The reactors of the Three Mile Island Nuclear Power Plant accident that occurred in the USA in 1979 are light water reactors.[309]

- **The advanced light water reactor:** The advanced light water reactor is a more technologically advanced version of the light water reactor. Compared to the light water reactor, the advanced light water reactor has advantages such as less investment, cheaper power generation, and a longer operating lifetime.[310] The first permission for the construction of advanced light water reactors in the world was obtained by the TEPCO company in Japan. There are two variants of the advanced light water reactor. The first is the advanced boiling water reactor. The second is the advanced pressurized water reactor.[311]

- **Pressurized heavy water reactor:** The pressurized heavy water reactor is one of the least common nuclear power reactors.[312] This reactor type was first developed in Canada.[313] Two types of heavy water[314] are used in the pressurized heavy water reactor. The first is used for cooling the heavy water

[307] B. Pershagen and M. Bowen, *Light Water Reactor*, 1.
[308] Geoffrey Rothwell and Bob van der Zwaan, "Are Light-water Reactor Energy Systems Sustainable?," *The Journal of Energy and Development* 29/1 (2003): 65-79, 66.
[309] Michael W. Golay and Neil E. Todreas, "Advanced Light-water Reactors," *Scientific American* 262/4 (1990): 82-89, 87.
[310] Kathleen Vaillancourt et al., "The Role of Nuclear Energy," 2298.
[311] "Status of Advanced Light Water Reactor Designs 2004," last updated date: 10 March 2022, https://www-pub.iaea.org/MTCD/Publications/PDF/te_1391_web.pdf
[312] Kathleen Vaillancourt et al., "The Role of Nuclear Energy," 2299.
[313] Belle R. Upadhyaya and Thomas W. Kerlin, *Dynamics and Control of Nuclear Reactors* (London: Academic Press, 2019).
[314] Heavy water is water with a high deuterium.

reactor. Second, heavy water is used to slow down the neutrons necessary for the fission chain reaction to occur while generating energy. Pressurized heavy water reactors, in which natural uranium is used instead of enriched uranium, constitute approximately 11% of nuclear energy reactors in the world.[315]

- **High-temperature gas-cooled reactor:** A high-temperature gas-cooled reactor is a nuclear fission reactor cooled by high-temperature helium gas. Gas-cooled reactors began to be developed in the 1950s, and high-temperature gas-cooled reactors began to be developed in the 1960s.[316] In this reactor type, the reactor core is exposed to high temperatures.[317] In the high-temperature gas-cooled reactor, the temperature reaches 750 degrees.

- **Nuclear fusion power reactors:** Nuclear fusion power reactors are a type of reactor under development. Such reactors are planned to be used in 2050.[318]

The third and final stage is decommissioning of nuclear power plants.[319] Nuclear power plants that have expired are decommissioned. Every nuclear power plant has an operating lifetime as in other power plants. The operating lifetime of nuclear power plants is 40 years.[320] However, the operating lifetime of nuclear power plants that are over 40 years old can be extended. The operating lifetime of nuclear power plants

---

315 "Are There Different Types of Nuclear Reactors?" last updated date: 12 March 2022, https://world-nuclear.org/nuclear-essentials/are-there-different-types-of-reactors.aspx

316 Zuoyi Zhang et al., "Design Aspects of the Chinese Modular High-temperature Gas-cooled Reactor HTR-PM," *Nuclear Engineering and Design* 236/5-6 (2006): 485-490, 485.

317 "Basic Studies in the Field of High-temperature Engineering," last updated date: 14 March 2022, https://www.oecd-nea.org/upload/docs/application/pdf/2019-12/5309-basic-studies.pdf, 96.

318 Kathleen Vaillancourt et al., "The Role of Nuclear Energy," 2299.

319 Benjamin K. Sovacool, "Valuing the Greenhouse Gas," 2951.

320 Lucas W. Davis, "Prospects for Nuclear Power," *Journal of Economic Perspectives* 26/1 (2012): 49-66, 59.

can be extended for another 20 years.[321] The operating lifetime of nuclear power plants, which do not have technical problems, can be extended for another 20 years for the second time. In short, nuclear power plants can be used for 80 years on average.

3. THE SCOPE OF NUCLEAR ENERGY

Technological developments in the field of nuclear energy are most commonly used in energy production. However, the use of nuclear technology is not limited to energy production. Nuclear technology is used in different fields and for different purposes. As a result of the rapid technological developments in the field of nuclear energy, the usage area is expanding in parallel. It is possible to examine the usage areas of nuclear technology under five headings. The first of these is the nuclear medicine sector. With the application of technology in the field of nuclear energy in the medical sector, a discipline called nuclear medicine[322] has emerged. Small-sized nuclear power plants in 56 countries on a global scale only produce isotope atoms for use in the medical and industrial sectors.[323]

Secondly, nuclear technology is used in the military field. There have been constant wars throughout human history. Despite international institutions established within the framework of modern human rights and democratic practices, wars still continue. The nuclear weapon is one of the most important war tools possessed by developed countries in weapon technology. The first nuclear weapons test was conducted in Mexico in 1945. In 1945,[324] right after the nuclear weapon test, the USA used nuclear weapons in the city of Hiroshima in the Second World War, causing thousands of people to die.[325] Nuclear weapons are produced from the

---

[321] Nico Bauera, Robert J. Brechab and Gunnar Luderer, "Economics of Nuclear Power and Climate Change Mitigation Policies," *Proceedings of the National Academy of Sciences* 109/42 (2012): 16805-16810, 16808.

[322] William D. Leslie, *Nuclear Medicine* (Boca Raton: CRC Press, 2003).

[323] Ian Hore-Lacy, *Nuclear Energy in the 21st Century* (London: World Nuclear University Press, 2010), 7.

[324] Steven L. Simon et al., "Estimated Radiation Doses Received by New Mexico Residents from the 1945 Trinity Nuclear Test," *Health Physics* 119/4 (2020): 428-477.

[325] Masao Tomonaga, "The Atomic Bombings of Hiroshima and Nagasaki: A Summary of the Human Consequences, 1945-2018, and Lessons for Homo Sapiens to end the Nuclear Weapon Age," *Journal for Peace and Nuclear Disarmament* 2/2 (2019): 491-517.

isotope Plutonium-239 (^{239}Pu, Pu-239), which is produced during energy production in nuclear power plants.[326]

Nuclear energy technology is also used in space exploration.[327] Nuclear energy technology has been used by both the USA and the Soviet Union since the 1960s.[328] Spaceships, which people use to go to space and do research around space, work with energy. Solar energy is widely used in spaceships. However, due to the insufficient solar energy, the Plutonium-238 radiant core produced in nuclear power plants is used for the operation of all technological devices in spaceships and at the same time, for heating. The energy obtained with plutonium batteries is also used in satellites and navigation instruments used in space studies.

Another area where nuclear technology is widely used is agricultural research and applications.[329] The world population has been increasing uncontrollably since the 1950s. Technological developments are prioritized in agricultural activities in order to meet the nutritional needs of the increasing population with limited resources in a stable manner. One of these technological developments is nuclear energy. Some countries use radiation technology to genetically enrich agricultural products such as grains, industrial crops, and fruit trees and to get more yields.[330]

4. THE HISTORY OF NUCLEAR ENERGY

Nuclear energy can be considered as a relatively new energy source. Nuclear energy is the newest energy source. Fossil fuels and renewable energy sources have been used for human needs for hundreds of years. However, nuclear energy first began to be used in the 1950s

---

[326] Siegfried S. Hecker, "Preventing Nuclear Weapon Proliferation as Nuclear Power Expands," *MRS bulletin* 33/4 (2008): 340-342, 340.

[327] Michael Maharikl and Baruch Fischhof, "The Risks of Using Nuclear Energy Sources in Space: Some Lay Activists' Perceptions," *Risk Analysis* 12/3 (1992): 383-392.

[328] Michael Maharika and Baruch Fischhoff, "Public Views of Using Nuclear Energy Sources in Space Missions," *Space Policy* 9/2 (1993): 99-108.

[329] Elaheh Ebrahimi Sarcheshmeh, Masoud Bijani and Hassan Sadighi, "Adoption Behavior Towards the Use of Nuclear Technology in Agriculture: A causal Analysis," *Technology in Society* 55 (2018): 175-182.

[330] Askari Mohammad Bagher et al., "Hydroelectric Energy Advantages," 102.

during the Cold War period.[331] The first commercial nuclear power plant in the world started to be used in the state of Pennsylvania, the USA in 1957. This nuclear power plant, which is a pressurized light water reactor, has a total power capacity of 60 MW(e).[332]

In the 1960s, the use of nuclear energy entered into a rapid increase in the USA. In order to meet the energy needs for its rapidly growing economy, there was a rapid development in terms of both quantity and quality in the use of nuclear energy in these periods in the USA. In the USA, which had only one nuclear power plant before 1960, 13 new nuclear power plants were built between 1960 and 1970. In this period, the construction process of 39 of them started. At the same time, the capacity of nuclear power plants was significantly improved during this period. In the 1960s, the energy production capacity of nuclear energy reactors with an electrical capacity of 90 megawatts was increased 8-10 times.[333]

One of the first countries to work in the field of nuclear energy is the Soviet Union. Nuclear energy studies in the Soviet Union emerged with the nuclear armament process in the Second World War. In 1942, "The Soviet atomic bomb project" was launched.[334] The first nuclear power plant in the Soviet Union was put into use on 27 June 1954 in the city of Obninsk, 107 kilometers from Moscow. This nuclear power plant with a capacity of 5,000 kilowatts was built in five years.[335] In 1985, the Soviet Union met 10% of its electricity needs from nuclear energy. Until 1990, the nuclear energy production of the Soviet Union continued to increase rapidly. The Soviet Union, which produced 170 million kilowatt-

---

[331] Michael Dittmar, "Nuclear Energy: Status and Future Limitations," *Energy* 37/1 (2012): 35-40, 35.

[332] W. J. Madia, "The 21st Century Nuclear Park: A Source of Energy, Water, Food, and Jobs," *International Journal of Nuclear Energy Science and Technology* 1/1 (2004): 7-19, 9.

[333] Roger E. Kasperson et al., "Public Opposition to Nuclear Energy: Retrospect and Prospect," *Science, Technology, & Human Values* 5/2 (1980): 11-23, 12.

[334] Martin Spence, "Soviet Power: Nuclear Energy in the USSR," *Capital & Class* 7/3 (1983): 87-96, 91.

[335] V. S. Emelyanov, "Nuclear Energy in the Soviet Union," *Bulletin of the Atomic Scientists* 27/9 (1971): 38-41, 41.

hours of electricity from nuclear energy until 1985, increased this to 360 million kilowatt-hours in 1990.[336]

China, which is the second largest nuclear energy producer after the USA, started producing nuclear energy in the 1960s. China's production of nuclear energy has emerged as a result of nuclear armament studies. China successfully tested its first nuclear bomb in 1964. Accelerating its work in the field of nuclear weapons in this period, China tested its first nuclear missile in 1966 and its first hydrogen bomb in 1967. It launched its first nuclear submarine in 1970. Nuclear science and the nuclear industry, which made rapid progress with the studies carried out in the field of nuclear weapons in these periods, constitute the foundations of China's nuclear energy infrastructure.[337]

In the 1970s and 1980s, the use of nuclear energy increased, especially in developed countries.[338] The main reason for this is the tendency to alternative energy sources to fossil fuels in order to ensure energy security due to the oil crisis in the 1970s.[339] However, due to the accidents in nuclear power plants in these periods, anti-nuclear energy began to rise. The first of these accidents, which affected people's attitudes towards nuclear energy and the energy policies of countries, was the Three Mile Island disaster that occurred in the USA in 1979. The second is the Chernobyl nuclear power plant accident that occurred in the Soviet Union in 1986. These accidents in nuclear power plant facilities have negatively affected people's attitudes towards nuclear energy.[340]

---

[336] A. Petrosyants, "The Soviet Union and the Development of Nuclear Power," *IAEA Bulletin* 28/3 (1986): 5-8, 6.

[337] Yun Zhou et al., "Is China Ready for Its Nuclear Expansion?," *Energy Policy* 39/2 (2011): 771-781, 771.

[338] Kathleen Vaillancourt et al., "The Role of Nuclear Energy," 2297.

[339] Peter W. Beck, "Nuclear Energy in the Twenty-first Century: Examination of a Contentious Subject," *Annual Review of Energy and the Environment* 24/1 (1999): 113-137.

[340] Cassandra L. Koerner, "Media, Fear, and Nuclear Energy: A Case Study," *The Social Science Journal* 51/2 (2014): 240-249.

Since 1990, the "renaissance" period has begun for nuclear energy.[341] With the introduction of new types of nuclear reactors, the interest in nuclear energy has increased. This situation continued until the Fukushima nuclear power plant accident in 2011. Immediately after the nuclear disaster in Japan, Germany and Switzerland decided to close their nuclear power plants gradually. Today, it is seen that the effects of Fukushima on energy policy are decreasing. Investment in nuclear energy tends to increase. It is possible to say that there has been a significant increase in nuclear energy projects, especially in China and South Korea.[342] In addition, countries that do not use nuclear energy have entered the process of building their first nuclear power plants.

5. NUCLEAR ENERGY WORLDWIDE

Since fossil fuels are found in certain geographical regions, energy control is under the sovereignty of a few countries. This situation carries serious risks, especially in terms of energy security. Because countries that do not have fossil fuels may have difficulty in meeting the energy demand in the domestic market in a stable manner. Unlike fossil fuels, it is possible to say that nuclear energy spreads over a wider geography. The main reason for this is that nuclear energy is based on technology transfer. All countries that have nuclear energy technology or countries that want to buy this technology can build nuclear power plants.

COUNTRY	NUMBER OF REACTORS	CAPACITY
USA	93	95523 MW
France	56	61370 MW
China	54	50789 MW
Russia	38	28578 MW
Japan	33	31679 MW
South Korea	24	23150 MW

341 Alessandra Dalla and Valle Claudia Furlan, "Diffusion of Nuclear Energy in Some Developing Countries," *Technological Forecasting and Social Change* 81 (2014): 143-153, 143.

342 Jochen Markard et al., "Destined for Decline? Examining Nuclear Energy from a Technological Innovation Systems Perspective," *Energy Research & Social Science* 67 (2020): 1-19.

COUNTRY	NUMBER OF REACTORS	CAPACITY
India	23	6885 MW
Canada	19	13624 MW
Ukraine	15	13107 MW
The United Kingdom	11	6848 MW

Table 10: Top 10 Countries with the Most Nuclear Energy Reactors [343]

As can be seen in Table 10, the world's largest nuclear energy producer is the USA.[344] A total of 95523 MW of energy is produced in the USA, where there are 93 nuclear energy reactors. The second country with the highest number of nuclear power reactors is France. The capacity of 56 nuclear energy reactors owned by France is 61370 MW. The third one is China. In China, where there are 54 nuclear energy reactors, 50789 MW of energy is produced. Russia, which ranks fourth, and Japan, which ranks fifth, have approximately similar nuclear energy capacities. While Russia, which has 38 nuclear energy reactors, produces 28578 MW of energy, this rate is 31679 MW in Japan, which has 33 nuclear energy reactors.[345]

The sixth country with the highest nuclear energy capacity is South Korea, another Asian country. 23150 MW of energy is produced in South Korea, where there are 24 nuclear energy reactors. The total capacity of 23 nuclear energy reactors in India, which is the second most populous country in the world after the People's Republic of China, is 6885 MW. The nuclear energy capacity of Canada, which ranks eighth, and Ukraine, which is ninth, are close to each other. In Canada, where there are 19 nuclear energy reactors, 13624 MW of energy is produced. The total capacity of Ukraine, where there are 15 nuclear energy reactors, is 13107 MW. The United Kingdom is in last place. The total capacity of 11 nuclear power reactors in the UK is 6848 MW.[346, 347]

---

[343] "Operational & Long-Term Shutdown Reactors," last updated date: 18 March 2022, https://pris.iaea.org/PRIS/WorldStatistics/OperationalReactorsByCountry.aspx

[344] Steven J. Zinkle and Gary S. Was, "Materials Challenges in Nuclear Energy," *Acta Materialia* 61/3 (2013): 735-758, 735.

[345] "Operational & Long-Term Shutdown Reactors"

[346] "Operational & Long-Term Shutdown Reactors"

[347] The countries with the most nuclear energy reactors in the world also constitute the largest economies in the world. Countries such as the USA, France, Russia, Japan, China, and India are the world's largest economies. The main raw material of the economy is

Nuclear power reactors are located in different regions of the world. The regions with the highest number of nuclear energy reactors in the world are the Far East and North America. In the Far East, where there are 114 nuclear energy reactors, the amount of energy produced from nuclear power plants is 108477 MW. In North America, where there are 112 nuclear energy reactors, 109147 MW of energy is produced. Although the number of reactors is less in North America, the amount of energy produced is higher, indicating that the capacity of nuclear power plants is higher. Another region with the highest number of nuclear energy reactors is the European continent. In Western Europe, where there are 99 nuclear energy reactors, 98454 MW of energy is produced.[348]

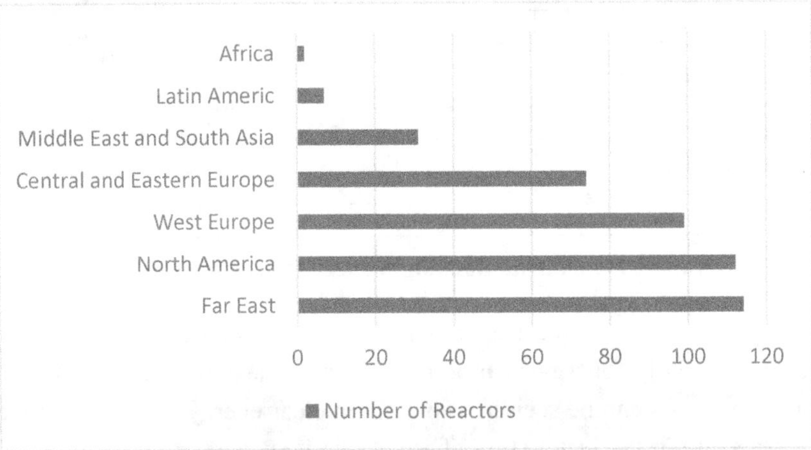

Table 11: Distribution of Nuclear Energy Reactors by Regions[349]

In Central and Eastern European countries, where there are 74 nuclear energy reactors, 54877 MW of energy is produced. There are 31 nuclear energy reactors in the Middle East and South Asian countries and 12732 MW of energy is produced. The regions with the least number of nuclear power reactors are Latin America and the African continent. In Latin America, where there are 7 nuclear energy reactors, 5077 MW of

---

energy. Mass production is not possible without energy. For this reason, countries with developed economies need more energy. Table 10 shows that nuclear energy has a critical role in meeting the energy needs of countries with the world's most developed economies.

348 "Operational & Long-Term Shutdown Reactors."

349 "Operational & Long-Term Shutdown Reactors."

energy is produced. On the African continent, there are only two nuclear energy reactors and their total capacity is 1860 MW.[350]

COUNTRY	NUMBER OF REACTORS	CAPACITY
China	14	14792 MW
India	6	4194 MW
Russia	4	3759 MW
South Korea	4	5360 MW
USA	2	2234 MW
The United Kingdom	2	3260 MW
Ukraine	2	2070 MW
Slovakia	2	880 MW
Japan	2	2653 MW
Pakistan	1	1014 MW
Iran	1	974 MW
France	1	1630 MW
Finland	1	1600 MW
Brazil	1	1340 MW
Belarus	1	1110 MW
Argentina	1	25 MW

Table 12: Countries Building Nuclear Energy Reactors[351]

The use of nuclear energy is in an increasing trend all over the world. Currently, energy is produced from nuclear power plants in 32 countries.[352] As can be seen in Table 12, nuclear energy reactors in many countries are in the installation phase. 45 new nuclear power reactors are being built in 16 countries using nuclear energy. This information shows us that the importance of nuclear energy is gradually increasing in countries where nuclear energy is used. In addition, the two countries that do not have nuclear power plants are in the process of building their first nuclear power reactors. The first of these countries is Bangladesh. In

---

[350] "Operational & Long-Term Shutdown Reactors."

[351] "Under Construction Reactors," last updated date: 3 April 2022, https://pris.iaea.org/PRIS/WorldStatistics/UnderConstructionReactorsByCountry.aspx

[352] "Nuclear Power in the World Today," last updated date: 3 April 2022, https://world-nuclear.org / information-library / current-and-future-generation /nuclear-power-in-the-world-today.aspx

Bangladesh, one of the developing economies, two nuclear energy reactors with a capacity of 2160 MW are being built.[353]

Another country that built the first nuclear power plant is Turkey. Nuclear energy has been taking place in Turkey's energy policy since the 1950s. Turkey's nuclear energy policy started with the Turkish Atomic Energy Agency, which was established in 1956. This was followed by the establishment of a 1 MW research reactor at the Nuclear Research and Training Center in Büyükçekmece in 1961.[354] In the second half of the 20th century, there were many attempts to construct nuclear power plants in Turkey. However, the policy of establishing a nuclear power plant could not be successful due to the political, economic, and international problems of that period.

The most important step Turkey took in the field of nuclear energy was realized in 2010. A Cooperation Agreement was signed between the Government of the Russian Federation and the Government of the Republic of Turkey, which envisages the construction of a nuclear power plant in the Akkuyu region in Mersin.[355] Four nuclear energy reactors, each with a power of 1200 MW, are being built. The type of these reactors is pressurized water reactors. Akkuyu Nuclear Power Plant, the foundations of which were laid in 2018, will be used in 2023. When it starts working at full capacity, it is estimated to meet approximately 10% of Turkey's electricity need with 35 billion kWh electricity production.[356] Turkey is also planning to build nuclear power plants in Sinop and Igneada regions.

6. CONCLUSION

In this section, the scope of nuclear energy, the types of nuclear power plants, the stages of energy production in nuclear power plants, the importance of nuclear energy, the history of nuclear energy, and the world's nuclear energy outlook are examined. First of all, what nuclear

---

353 "Under Construction Reactors."

354 Kadir Temurçin and Alpaslan Aliağaoğlu, "Nükleer Enerji ve Tartışmalar," 32.

355 "Akkuyu NGS İnşaat Projesi," last updated date: 4 April 2022, http://www.akkuyu.com/

356 "Akkuyu NGS İnşaat Projesi."

energy means, its working principles, and the scope of nuclear power plants are examined. Nuclear energy is based on the conversion of energy produced by splitting uranium by colliding with neutrons. For this reason, the main fuel used in nuclear power plants is uranium. There are five basic types of nuclear power plants. They are as follows: (1) light water reactors, (2) advanced light water reactors, (3) pressurized heavy water reactors, (4) high temperature gas cooled reactors, and (5) high temperature gas cooled reactor.

Secondly, in which sectors nuclear energy technology is used is examined. Nuclear energy is most commonly used for electricity generation. However, the use of technology in this field is not limited to the energy sector. It is also used in different fields. The first of these is the arms industry. Nuclear weapons are produced from the isotopes of Uranium-235 and Plutonium-239, which are produced during power generation in nuclear power plants. As a matter of fact, the starting point of nuclear energy technology is based on the nuclear armament process in the Second World War. Another usage area of nuclear energy technology is space studies. Nuclear energy is used in the production of energy in space vehicles due to the inadequacy of renewable energy sources. Nuclear energy technology is also widely used in medical studies and agricultural activities.

Thirdly, how nuclear energy emerged and developed in the historical process has been examined. The emergence of nuclear energy is based on the armament process for military purposes. With the efforts to produce nuclear weapons in the Second World War, nuclear energy technology began to emerge. During the Cold War period, nuclear energy technology showed rapid development with the competition between the USA and the Soviet Union. The first commercial nuclear power plant was opened in the Soviet Union in 1954. Immediately after, nuclear power plants started to be established in the USA. China also participated in this competition in the 1960s. The use of nuclear energy has started to become widespread in developing countries since the 1970s. However, the use of nuclear energy has slowed down due to the Three Mile Island accident in 1979 and the Chernobyl nuclear power plant accident in 1986. With the use of new types of nuclear energy reactors in the 1990s, the interest in nuclear energy tended to increase and this situation continued until the Fukushima disaster that occurred in 2011.

Finally, the world's nuclear energy outlook was analyzed on a country-by-country basis. 32 countries around the world produce energy in nuclear power plants. The first two countries with the highest nuclear energy production are the USA and France, respectively. China and Russia are next. Considering the regions, the geographical areas with the highest number of nuclear energy reactors in the world are the Far East and North America. The Far East, where there are 114 nuclear energy reactors, ranks first among the regions with nuclear power plants in the world. Although nuclear energy investments have decreased in the world after the Fukushima nuclear power plant accident in 2011, the construction of new nuclear power plants continues. 45 new nuclear power reactors are being built in 16 countries using nuclear energy. In addition, Turkey and Bangladesh are in the process of establishing their first nuclear energy reactors.

CHAPTER IV:

ADVANTAGES OF NUCLEAR ENERGY

1. INTRODUCTION

Nuclear energy is one of the most important energy sources in the world. The demand for nuclear energy tends to increase, albeit partially, all over the world. Currently, 50 nuclear energy reactors are under construction.[357] One of the most important reasons for this is the desire to reduce dependence on fossil fuels. Nuclear energy is one of the energy sources that has emerged as an alternative to fossil fuels, which are at risk of depletion and have high environmental costs. Nuclear energy is at the center of the energy policy of many countries due to its significant advantages in terms of the environment and society. However, creating an alternative energy source to fossil fuels is not the only reason behind the demand for nuclear energy. Nuclear energy has significant advantages compared to other energy sources.

Nuclear energy has many advantages. It is at the center of the energy policy of many countries due to its significant advantages in terms of environment, society, and economy. Nuclear energy has a key role in meeting energy needs, especially in economically developed countries. For example, France, one of the largest economies in the world, meets approximately 70% of its energy needs from nuclear energy. Similarly, there are 60 nuclear power plants in the USA, the world's largest economy. Nuclear energy use is also widespread in other major economies such as Japan, China, and Russia.

---

[357] "Nuclear Power in the World Today," last updated date: 17 January 2022, https://world-nuclear.org/information-library/current-and-future-generation/nuclear-power-in-the-world -today.aspx

The main purpose of this section is to discuss the advantages of nuclear energy in terms of environment, public health, and economy. The advantages of nuclear energy discussed in this section are as follows; (1) the long operating lifetime of nuclear power plants, (2) low greenhouse gas emissions caused by nuclear power plants, (3) nuclear power plants do not release pollutants into the air, (4) high efficiency from nuclear energy, (5) nuclear energy power plants provide energy security, (6) nuclear energy contributes to the increase in employment by creating job opportunities at the regional level, (7) requires a low level of land use, and (8) contributes to reducing foreign dependency on energy.

2. ADVANTAGES OF NUCLEAR ENERGY

Nuclear energy has many advantages. This section focuses on the most important advantages of nuclear energy. This section examines eight key advantages of nuclear energy.

2.1. The Long Lifetime of Nuclear Power Plants

Every physical place built by humans in nature has a lifespan in the world. All energy power plants established for the purpose of obtaining energy also have a lifetime. The service life of power plants varies according to the type of power plants. The type of power plant with the longest usage period is the hydroelectric power plant. The average lifetime of hydroelectric power plants built on the river and used only for electricity generation is 80 years. After hydroelectric power plants, the type of power plant with the longest lifetime is the nuclear power plant. With today's technology, the average usage time of nuclear power plants is 60 years.

The third type of energy power plant with the longest lifespan is coal power plants and geothermal power plants. Coal-fired power plants are power plants that work with steam power using coal. A geothermal power plant is a power plant used to convert the energy obtained from geothermal resources into electricity. The average usage period of thermal and geothermal power plants is 40 years. The type of power plant with the shortest lifetime is those used to generate energy from renewable energy sources. The average lifetime of wind and solar power plants is 25 years.

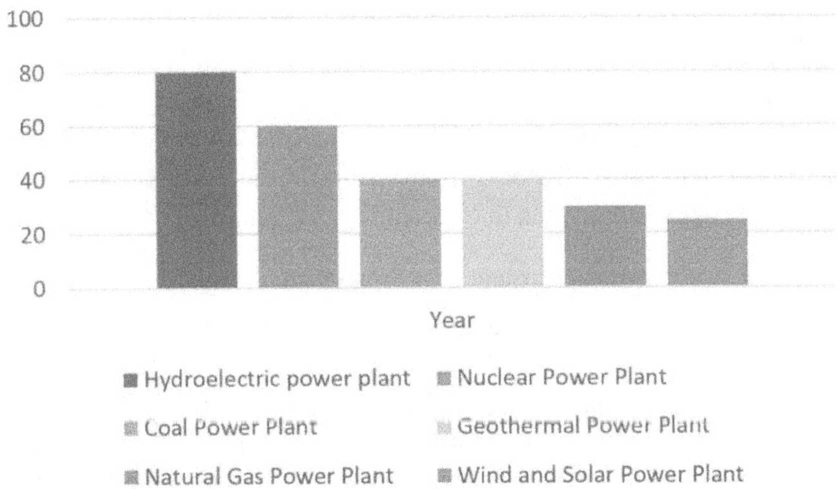

Table 13: Average Usage Period of Power Plants[358]

On the other hand, it should be noted that in parallel with the development of nuclear energy technology, the average lifetime of nuclear power plants is increasing. For example, until 2000, the operating lifetime of nuclear power plants varies between 20 and 40 years. Today, however, the license for use of nuclear power plants can be extended from 40 to 60 years.[359] In addition, it is possible to extend the lifetime of nuclear power plants by 20 more years for the second time. From this point of view, the lifetime of nuclear power plants can reach 80 years. For this reason, nuclear power plants can be defined as the power plants with the longest lifetime together with hydroelectric power plants.

The long operating lifetime of power plants is of strategic importance in terms of both the economy and the environment. First, there is an economic cost to build every power plant. Reconstruction of power plants with short lifetimes at regular intervals results in an increase in economic costs. Since the cost of power plants will increase, energy prices will increase in parallel. This situation poses a risk to ensuring energy

---

358 "Projected Costs of Generating Electricity," last updated date: 18 January 2022, https://iea.blob.core.windows.net/assets/ae17da3d-e8a5-4163-a3ec-2e6fb0b5677d/Projected-Costs-of-Generating-Electricity-2020.pdf, 36.

359 Todd Allen et al., "Materials Challenges for Nuclear Systems," *Materials Today* 13/12 (2010): 14-23, 14.

security. Second, the construction of power plants is possible with the use of natural resources. All resources in nature are available in limited quantities. The use of natural resources for power plants can increase the risk of resource depletion. For this reason, the use of power plants, which can obtain long-term efficiency from natural resources, may provide a more advantageous situation in terms of the environment.

2.2. Low Greenhouse Gas Emissions

It is generally accepted all over the world that the most important environmental problem of the 21st century is climate change. Since climate change is a global problem, all countries struggle with environmental problems caused by climate change. The most important factor causing climate change is the increase in the average temperature of the earth. The main reason for this situation, which is called global warming, is the greenhouse gases released into the atmosphere. Carbon dioxide, methane, nitrogen oxide, chlorofluorocarbons, and carbon tetrachloride (CCI_4) are the most common greenhouse gases.[360] These gases, which absorb and emit infrared radiation, cause an increase in the average temperature of the earth, triggering climate change.

The sources of greenhouse gases differ from each other. For example, 63% of carbon dioxide, which constitutes 70% of greenhouse gases, originates from fossil fuels.[361] However, 45% of methane gas, which is the most common greenhouse gas after carbon dioxide and accounts for 30% of the increase in global average temperature,[362] originates from agricultural activities and 25% comes from animals.[363] There are two main energy sources in reducing or preventing environmental problems caused by fossil fuels.[364] These are nuclear

---

[360] T. R. Watson et al., "Greenhouse Gases: Sources and Sinks," *Climate Change* 92 (1992): 25-46, 29.

[361] Jos G. Olivier et al., "Recent Trends in Global Greenhouse Gas Emissions: Regional Trends 1970–2000 and Spatial Distributionof Key Sources in 2000," *Environmental Sciences* 2(/2-3 (2005): 81-99, 82-83.

[362] "World Energy Outlook 2021," last updated date: 19 January 2022, https://iea. Blob .core. windows.net / assets / 4ed140c1-c3f3-4fd9-acae-789a4e14a23c/WorldEnergy Outlook2021. pdf, 41.

[363] Jos G. Olivier et al., "Recent Trends," 83.

[364] Aviel Verbruggen, "Renewable and Nuclear Power," 4046

energy and renewable energy, which do not cause greenhouse gas emissions.

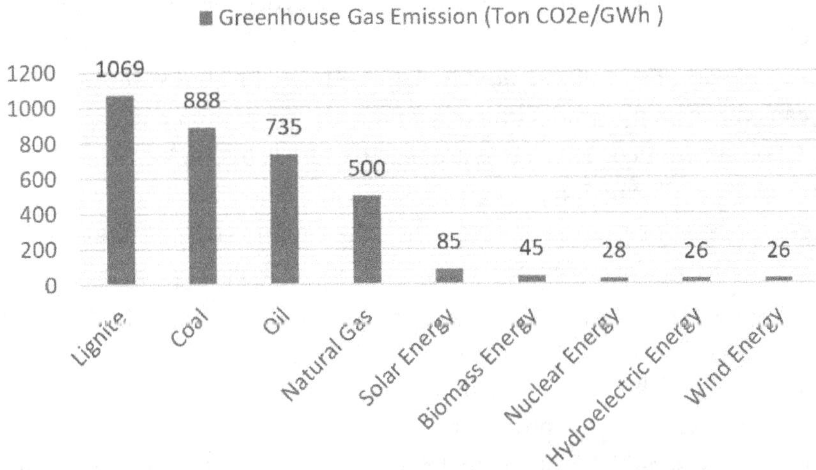

Table 14: Greenhouse Gases Emission Amount of Energy Sources[365]

As seen in Table 14, the energy source with the highest greenhouse gas emissions is fossil fuels. The first of these is lignite. Lignite causes 1069 CO_2e[366] emissions in 1 GWh energy production. Coal is in second place, which is one of the fossil fuels that causes the most greenhouse gas emissions. Coal triggers 888 CO_2e emissions in 1 GWh energy production. The third and fourth energy sources that cause the most greenhouse gas emissions are oil and natural gas, respectively. Oil causes 735 CO_2e in 1 GWh energy production, and natural gas causes 500 CO_2e emissions. In the light of this information, it can be noted that natural gas has the lowest greenhouse gas emissions among fossil fuels.

Renewable energy sources also cause greenhouse gas emissions. Among the renewable energy sources, it is solar energy that causes the most greenhouse gas emissions. In 1 GWh energy production from solar energy, 85 CO_2e is emitted. However, it is possible to say that this rate is at very low levels when compared to fossil fuels. Biomass energy is in

[365] "Comparison of Lifecycle Greenhouse Gas Emissions of Various Electricity Generation Sources," last updated date: 22 January 2022, http://www.world-nuclear.org/uploaded Files/org/WNA/Publications/Working_Group_Reports/comparison_of_lifecycle.pdf
[366] CO_2e refers to the carbon dioxide equivalents of greenhouse gases.

second place after solar energy. In 1 MWh energy production from biomass energy, 45 CO_2e is released.

The energy sources with the lowest greenhouse gas emissions are nuclear energy, hydroelectric energy, and wind energy. The greenhouse gas emission rates of these three energy sources are approximately close to each other. In the production of 1 MWh of energy from nuclear energy, 28 CO_2e are emitted. This rate is around 26 CO_2e in hydroelectric and wind energy. This information shows us that the most environmentally friendly energy sources that can be used to reduce the emission of greenhouse gases in the fight against climate change are nuclear energy, hydroelectric energy, and wind energy.

If the world did not use nuclear power plants, CO_2 emissions would be higher by about 2.5 billion tons per year.[367] A 1% increase in the use of nuclear energy provides a 0.477% reduction in emissions.[368] However, it should be noted that nuclear energy indirectly causes the emission of greenhouse gases.[369] Energy is used in all nuclear energy-related processes, such as uranium mining, construction of nuclear power plants, operation of nuclear power reactors, decommissioning of nuclear power plants, fuel reprocessing, nuclear waste disposal, and nuclear waste storage.[370] Since this energy is met from fossil fuels, nuclear energy indirectly causes the emission of greenhouse gases.

2.3. Clean Air

Access to clean air is recognized as a basic human right. Because clean air is one of the minimum conditions necessary for people to continue their lives in a healthy way and to protect their physical and mental health. Clean air is as important to humans as food and water.

---

[367] Achilles Adamantiades and Ioannis Kessides, "Nuclear Power for Sustainable Development: Current Status and Future Prospects," *Energy Policy* 37/12 (2009): 5149-5166.

[368] Nicholas Apergis et al., "On the Causal Dynamics Between Emissions, Nuclear Energy, Renewable Energy, and Economic Growth," *Ecological Economics* 69/11 (2010): 2255-2260, 2259.

[369] Vasilis M. Fthenakis and Hyung Chul Kim, "Greenhouse Gas Emissions from Solar Electric and Nuclear Power: A life Cycle Study," *Energy Policy* 35/4 (2007): 2549-2557.

[370] Manfred Lenzen, "Life Cycle Energy and Greenhouse Gas Emissions of nuclear Energy: A Review," *Energy Conversion and Management* 49/8 (2008): 2178-2199.

However, due to nature's own cycle and human actions, the natural components of the air are deteriorated and polluted. As a matter of fact, one of the most important environmental problems of the 21st century is air pollution. In the period we live in, nine out of ten people live in polluted air conditions.[371] This situation carries serious risks for life. Due to air pollution, millions of people experience health problems and die as a result. According to the World Health Organization, approximately 7 million people die every year due to air pollution.[372]

There are many causes of air pollution. However, one of the most important reasons among these is excessive fossil fuel consumption. Especially with the widespread use of coal and oil, not only the emission of greenhouse gases has increased, but also the emission of pollutants that damage the natural components of the air has increased rapidly and continues to increase. Fossil fuels, which have been used to meet energy needs for many years, trigger air pollution by causing the release of pollutants into the atmosphere. Coal is the most dangerous of all fossil fuels. Thermal power plants, where coal is burned and converted into energy, are responsible for 95% of $PM_{2.5}$ particles caused by fossil fuel power plants.[373] $PM_{2.5}$ particle is one of the main causes of air pollution. Likewise, the use of oil and natural gas causes the release of particulate matter (PM) and toxic substances (air toxics) into the air.[374] Therefore, in order to prevent air pollution, there is a need to use energy sources that do not cause the release of pollutants.

---

[371] "9 Out of 10 People Worldwide Breathe Polluted Air," last updated date: 25 January 2022, https://www.who.int/news/item/02-05-2018-9-out-of-10-people-worldwide-breathe-polluted-air-but-more-countries-are-taking-action#:~:text=Air%20pollution%20levels%20 remain%20dangerously,outdoor)%20and%20household%20air%20pollution.

[372] "New WHO Global Air Quality Guidelines Aim to Save Millions of Lives from Air Pollution," last updated date: 26 January 2022, https://www.who.int/news/item/22-09-2021-new-who-global-air-quality-guidelines-aim-to-save-millions-of-lives-from-air-pollution

[373] Xinming Du et al., "Transboundary Air Pollution from Coal-fired Power Generation," *Journal of Environmental Management* 270 (2020): 1-9, 2.

[374] Tammy M. Thompson et al., "Modeling to Evaluate Contribution of Oil and Gas Emissions to Air Pollution," *Journal of the Air & Waste Management Association* 67/4 (2017): 445-461.

In order to prevent air pollution, the world has been tending towards clean energy sources from fossil fuels in recent years.[375] One of the energy sources considered as a clean energy source in preventing air pollution is nuclear energy. Because while energy is produced in nuclear power plants, no pollutant is released into the air,[376] that is, it does not cause air pollution. For this reason, it is possible to say that nuclear energy has a key role in preventing air pollution caused by fossil fuels and reducing its effects.

2.4. Fuel Efficiency

All resources in nature are available in limited quantities. There is a risk of depletion of both underground and surface resources. The same is true for energy resources. For example, it is estimated that fossil fuels will run out on average within 100 years. The risk of depletion is not limited to fossil fuels. Contrary to what is commonly known, renewable energy sources also have a risk of depletion. Because the technological tools used in renewable energy sources are obtained by converting the limited amount of resources in nature. For all these reasons, it is necessary to use energy sources where the highest efficiency can be obtained from the energy source. In order to meet the energy needs in a sustainable way and to ensure the security of the energy needs of future generations, it is more rational in terms of energy policies to prefer energy sources with high fuel efficiency.

The fuel efficiency of energy sources differs significantly from each other. Because different types of fuels are used in each energy source. The most fuel-efficient energy source is renewable energy. Renewable energy sources, unlike fossil fuels, are based on constantly self-renewing sources such as sun, wind, waves, and water. It is possible to define these sources as the basic fuels of renewable energy. Since these resources are constantly renewing themselves, the fuel converted into energy in renewable energy sources does not run out or decrease.

---

[375] Ming-Jia Li et al., “The Development Technology and Applications of Supercritical CO_2 power Cycle in Nuclear Energy, Solar Energy and Other Energy İndustries,” *Applied Thermal Engineering* 126 (2017): 255-275, 256.

[376] Achilles Adamantiades and Ioannis Kessides, “Nuclear Power for Sustainable,” 5151.

After renewable energy sources, the energy source with the highest fuel efficiency is nuclear energy. It is possible to obtain high energy efficiency with low fuel amount in nuclear power plants. As mentioned before, the main fuel used in energy production in nuclear power plants is uranium. The amount of energy obtained from uranium is quite high compared to fossil fuels. The energy to be obtained from 1 kilogram of uranium in a nuclear reactor is equivalent to the energy produced by burning approximately 3000 tons of coal.[377] To explain with a more concrete example, it is possible to produce 8 kilowatts of energy with 1 kilo of coal, 12 kilowatts of energy with 1 kilo of oil, while it is possible to produce 24,000,000 kilowatts of energy with 1 kilo of uranium-235.[378] These examples show that the fuel efficiency in nuclear energy is considerably higher than the fuel efficiency in fossil fuels.

2.5. Energy Security

Energy security is a multidimensional concept.[379] In its most common definition, energy security means uninterrupted energy resources. It is the constant accessibility of energy within the framework of the supply-demand relationship.[380] Providing stable energy is one of the most important priority issues of the 21st century. Energy security is at the center of the energy policy of countries.[381] Because energy is one of the most important needs that determine the quality of life in a country. People constantly need and consume energy. Energy production and distribution must be met continuously, that is, energy access must be ensured in a stable manner. Therefore, energy security is one of the issues that both states and individuals prioritize.[382]

---

[377] Walter C. Patterson, *Nuclear Power* (Harmondsworth: Penguin, 1976), 9.

[378] "Fuel Comparison," last updated date: 25 January 2022, https://www.euro nuclear.org/ glossary/fuel-comparison/

[379] Bert Kruyt et al., "Indicators for Energy Security," *Energy Policy* 37/6 (2009): 2166-2181.

[380] J. Bielecki, "Energy Security: Is the Wolf at the Door?," *The Quarterly Review of Economics and Finance* 42/2 (2002): 235-250.

[381] Christian Winzer, "Conceptualizing Energy Security," *Energy Policy* 46 (2012); 36-48, 36.

[382] Roland Dannreuther, *Energy Security* (Cambridge: John Wiley & Sons, 2017).

Renewable energy is the most criticized energy source in terms of providing energy without interruption. Renewable energy sources are directly affected by climatic conditions. It is not always possible to get the same efficiency from renewable energy sources. For example, in Turkey, where there are 4 seasons, the efficiency obtained from solar energy in the summer season and the efficiency obtained in the winter season are completely different from each other, in other words, the efficiency changes according to the season.[383] Similarly, while high efficiency is obtained from solar energy during the day, efficiency cannot be obtained after sunset due to the absence of the sun. For this reason, it is difficult to produce energy regularly and consistently from renewable energy sources with the current technology. One of the energy sources that has the most important advantage in ensuring energy security is nuclear energy.[384]

When we look at it from the point of view of the economic discipline, energy security also means that energy is available at an affordable price.[385] Especially after the 1973 Oil Crisis, with the high increase in energy prices, many countries had serious problems in accessing energy, and this situation made energy security more debatable.[386, 387] Since energy is at the center of our lives in the period we live in, it is critical that energy prices be affordable in access to energy. If energy prices are high, only economically strong groups can access energy. However, the economic power of individuals in society can differ significantly from each other.

According to the United Nations Development Programme, 1.3 billion people, or 21.7% of the world's population, live in multifaceted poverty.[388] Economically disadvantaged groups also need the energy to sustain their lives in a healthy way. For example, it is a necessity to use resources such as natural gas and coal to meet the need for heating in

---

[383] L. Lakatos, G. Hevessy and J. Kovács, "Advantages and Disadvantages of Solar Energy and Wind-power Utilization," *World Futures* 67/6 (2011): 395-408, 395-396.

[384] Kathleen Vaillancourt et al., "The Role of Nuclear Energy," 2297.

[385] Christian Winzer, "Conceptualizing Energy Security," 36.

[386] Douglas R. Bohi and Michael A. Toman, *The Economics of Energy Security* (Massachusetts: Kluwer Academic Publishers, 2012), 3.

[387] Daniel Yergin, "Ensuring Energy Security," *Foreign Affairs* 85/2 (2006): 69-82.

[388] "The 2021 Global Multidimensional Poverty Index (MPI)," last updated date: 28 January 2022, http://hdr.undp.org/sites/default/files/2021_mpi_report_en.pdf, 1.

winter. However, in the case of high energy prices, it may be difficult for the poor to access energy. In this case, only those who are economically able can access energy. Regardless of their ethnic origin, belief, ideology, social status, and economic status, energy should be accessible to everyone, regardless of their economic status.

Energy security also means the diversity of energy sources.[389] Meeting the energy need of a country only or predominantly from a single source poses a risk in ensuring energy security. If the energy need is met from different sources, the gap that will occur due to the problem in one of the energy sources can be compensated with the other energy source. For example, a country that meets its energy needs only from fossil fuels may face serious problems in accessing energy and high energy costs if oil and natural gas prices rise in the international market due to wars and similar crises. For this reason, meeting the energy need by different energy sources, that is, diversification of energy sources, has strategic importance in order to ensure energy security.

Energy security means that the sources of energy supply are diverse.[390] Meeting the energy supply from a single source poses a risk to energy security. Because if there is a problem with the energy supply or if the energy supply is in trouble, risks may arise in accessing energy. For example, if a country meets its energy needs by importing from a single country, the problem of not accessing energy in the supplier country may arise in cases of war and similar situations. In this case, since there is no alternative supplier, the energy importing country may face serious problems in accessing energy. Therefore, the diversification of the sources of energy supply is accepted as one of the minimum conditions to ensure energy security.[391]

---

389 Abdelrahman Azzuni and Christian Breyer, "Energy Security and Energy Storage Technologies," *Energy Procedia* 155 (2018): 237-258, 238.

390 Beng Wah Ang, Wei Lim Choong and Tsan Sheng Ng, "Energy Security: Definitions, Dimensions and Indexes," *Renewable and Sustainable Energy Reviews* 42 (2015): 1077-1093.

391 Turkey can be cited as an example for the diversification of sources of energy supply. Turkey supplies a significant portion of fossil fuels by importing. But it does not make this supply from a single country. Some of the countries where Turkey supplies fossil fuels are as follows; Russia, Iran, Azerbaijan, Qatar, Nigeria and Algeria.

Energy security also includes environmental security.[392] Energy resources have indirect costs. One of these costs is the damage caused by energy sources to the environment. Environmental pollution occurs during the direct use of some energy sources. In particular, the conversion of fossil fuels into energy has devastating effects on the physical and biological elements of the environment. Due to the environmental effects of fossil fuels, the cost is also high, and it poses a risk in terms of safety. For this reason, it is very important to choose energy sources with low environmental costs in terms of energy security.

One of the energy sources that does not directly harm the environment while producing energy is nuclear energy. During energy production in nuclear power plants, pollutants are not released into the atmosphere. This contributes positively to reducing the environmental cost of energy resources. However, it should be noted that radioactive contamination that will occur in the event of a possible accident may cause more environmental costs than other energy sources. In addition, if the nuclear wastes generated during energy production in nuclear power plants cannot be stored safely, the environmental cost may increase even more. For all these reasons, it is necessary to prioritize security measures in nuclear energy in order to ensure environmental energy security.

2.6. Economic Contribution

One of the most important problems in the countries where the neoliberal economy is implemented is unemployment. Unemployment can be defined as the inability of an individual to find a job despite wanting to work in a society. According to the International Labor Organization, the estimated unemployment rate worldwide in 2022 is 5.9%. However, the unemployment rate varies considerably from country to country. For example, while the estimated unemployment rate for those over the age of 15 in Afghanistan in 2022 is 18.5%, this rate is 0.1% in Qatar.[393]

---

[392] Adam Simpson, "The Environment–Energy Security Nexus: Critical Analysis of an Energy 'Love Triangle'in Southeast Asia," *Third World Quarterly* 28/3 (2007): 539-554.

[393] "Unemployment Rate," last updated date: 28 January 2022, https://ilostat.ilo.org/data/#

In countries with high unemployment, poverty and deprivation also increase.[394] Because the purchasing power of the household starts to decrease. In order for people to lead their lives in a healthy way and to live in prosperity, they need to earn money, that is, to work. For this reason, it is essential to increase job opportunities in a region or country in order to improve the living standard of the society and for individuals to reach a more prosperous life. One of the most important sectors contributing to increasing employment is energy. Millions of people are employed in the realization of processes such as the production, operation, and distribution of the energy we use in many stages of our lives.

One of the energy sources that creates the most employment opportunities is nuclear energy. Nuclear energy engineers are not only employed in the nuclear energy sector. Some of the occupational groups employed in the nuclear energy sector are as follows; chemists, reactor operators, safety and environmental impact specialists, chemical engineers, communicators, radiation protection specialists, civil engineers, health physicists, mechanical engineers, scientists, lawyers, heavy equipment operators, project manager, accountant, cybersecurity specialists, carpenter, electrician, plumber, sheet metal workers, welder, subject matter expert, policy analyst, and financial manager.[395] In short, the nuclear energy sector creates employment opportunities for people from many different occupational groups.

Between 500 and 800 workers are employed at each nuclear power plant. This number varies according to the size of nuclear power plants. More than one generation has the opportunity to work in nuclear power plants with an average life of 80 years. The area where the most employment is provided in the nuclear energy sector is the installation of nuclear power plants. Approximately 7 thousand workers are employed in the construction of nuclear energy reactors.[396] The construction process of

---

394 P. Saunders, "The Direct and Indirect Effects of Unemployment on Poverty and Inequality," *Australian Journal of Labour Economics* 5/4 (2002): 507-529.

395 "Quick Facts on Nuclear Industry Jobs," last updated date: 30 January 2022, https://www.nei.org/advantages/jobs#:~:text=A%20single%20nuclear%20power%20plant,type%20of%20energy%20generation%20facility.&text=The%20U.S.%20nuclear%20energy%20sector,when%20you%20include%20secondary%20jobs.

396 "Quick Facts on Nuclear Industry Jobs."

nuclear power plants takes 5-6 years on average. 80% of nuclear power plants are installed in less than 120 months.[397] During this time, thousands of people are employed in the installation of nuclear power plants. The number of people employed in the nuclear energy sector worldwide is around 0.8 million.[398]

One of the most comprehensive studies on the job opportunities created by the nuclear energy sector has been carried out by the Nuclear Energy Agency and The International Atomic Energy Agency. According to the report prepared by these two institutions, the information about the number of people employed in the nuclear power plant with only 1000 MWe (net) advanced light water unit is as follows; (1) 600 administrative, operation and maintenance, and permanent contract personnel are employed annually during the 50-year operational period of the nuclear power plant, (2) 500 persons annually during the 10-year decommissioning period of the nuclear power plant, and (3) 80 persons are employed for the management of nuclear wastes.[399]

The number of people employed in the nuclear energy sector differs from country to country. Because the number and capacity of nuclear power plants differ from each other in countries using nuclear energy. Hundreds of thousands of people are employed in the nuclear energy sector in the USA, which is the world's largest nuclear energy producer. The nuclear energy sector employs more than 100,000 full-time employees in the United States alone. When the secondary works related to nuclear energy are included, this number approaches around 400,000.[400] In South Korea, this figure is relatively much less. In 2009, the nuclear energy sector in South Korea created employment for 29,400 people. On the other hand, 36 thousand 700 people were employed in secondary works related to the nuclear energy sector. In France, the number of people

---

[397] João ML Moreira, Bruno Gallinaro andPedro Carajilescov, "Construction Time of PWRs," *Energy Policy* 55 (2013): 531-542, 541.
[398] Sandeep Pai et al., "Meeting Well-below 2 C Target Would Increase Energy Sector Jobs Globally," *One Earth* 4/7 (2021): 1026-1036, 1027.
[399] Victoria Alexeeva, et al., "Measuring Employment Generated by the Nuclear Power Sector," No. NEA--7204. Organisation for Economic Co-Operation and Development, 2018, 12.
[400] "Quick Facts on Nuclear Industry Jobs."

directly employed in the nuclear energy sector is 125,000. The number of people employed in secondary jobs related to the nuclear energy sector is 114,000.[401]

Figure 1: Mass Housing for Employees Near the Chernobyl Nuclear Power Plant[402]

Since hundreds of people are employed in the operation of nuclear power plants, living spaces are created around the nuclear power plants in order to meet the housing needs of the employees and their families. For example, the photograph in Figure 1 shows the housing estates around the Chernobyl Nuclear Power Plant. Hundreds of similar housing estates were built only in order to meet the accommodation needs of the employees and their families at the Chernobyl Nuclear Power Plant. After the accident that occurred in 1986, these residences were abandoned. It is possible to see similar housing estates around all nuclear power plants in the world. Because nuclear power plants employ thousands of people.

The energy source that creates employment opportunities is not only nuclear energy. Other energy sources also contribute directly and indirectly to the employment of millions of people around the world. 18 million people are working in the energy sector in the world. The Fossil energy sector constitutes the biggest among these. 12.6 million people in the world are employed in the fossil energy sector. The renewable energy sector comes in second place. The renewable energy sector offers employment opportunities to millions of people in many countries of the

[401] Victoria Alexeeva, et al., "Measuring Employment," 12.
[402] This photo was taken by the author during his doctoral studies in 2017.

world.[403, 404] The number of people employed in the field of renewable energy in the world is 4.6 million.[405] However, it is thought that the number of people to be employed in this field will increase due to the increase in the use of renewable energy sources. For example, according to a study, it is estimated that the renewable energy sector will contribute to the employment of 4 million full-time employees by 2030 in the USA alone.[406]

In summary, one of the important advantages of nuclear energy is its positive contribution to the reduction of unemployment. The nuclear energy sector offers job opportunities to many people in different professions around the world. It, directly and indirectly, contributes to the full-time employment of hundreds of thousands of people from different occupational groups at every stage of nuclear energy. With the increase in the use of nuclear energy, the number of people to be employed is likely to increase as well. However, this advantage in the field of the economy does not only belong to the nuclear energy sector. Millions of people are employed in different occupational groups in the fossil fuel and renewable energy sectors.

2.7. Reducing External Dependency on Energy

As the countries' economies grow and their population increases, their need for energy increases in parallel. However, fossil fuels, which meet a significant part of the world's energy needs, are found in nature in limited quantities, that is, they decrease as they are used. Another disadvantage of fossil fuels is that they are not available in sufficient quantities in every country and geography. Fossil fuels are a type of energy source found in certain geographical regions. For example, there are 1732.4 million barrels of oil detected in the world as of 2020. 48.3% of

---

[403] Maria Isabel Blanco and Glória Rodrigues, "Direct Employment in the Wind Energy Sector: An EU Study," *Energy Policy* 37/8 (2009): 2847-2857.

[404] Ulrike Lehr et al., "Renewable Energy and Employment in Germany," *Energy Policy* 36/1 (2008): 108-117.

[405] Sandeep Pai et al., "Meeting Well-below," 1027.

[406] Max Wei, Shana Patadia and Daniel M. Kammen, "Putting Renewables and Energy Efficiency to Work: How Many Jobs Can the Clean Energy Industry Generate in the US?," *Energy Policy* 38/2 (2010): 919-931, 919.

this oil is found in the Middle East and Africa regions.[407] There is a similar problem with natural gas. 41% of the world's natural gas reserves are located in the Middle East region.[408] Since a significant part of fossil fuels is located in a limited geography, many countries do not have these resources, so they have to import them, that is, they are dependent on foreign sources for energy.

Oil Importing Countries	Amount of Imported Oil
China	459 Mt[409]
USA	292 Mt
India	226 Mt
South Korea	151 Mt
Japan	151 Mt
Germany	85 Mt
Spain	67 Mt
Italy	63 Mt

Table 15: Top Oil Importing Countries in 2018 [410]

As seen in Table 15, the world's largest economies and countries with the largest population export oil to meet their energy needs. The People's Republic of China leads these countries. China, the world's most populous country and the world's second largest economy, exported 459 metric tons of oil in 2018. The USA comes right after. In 2018, the USA exported 292 metric tons of oil. India, which has the second largest population in the world, ranks third with 226 metric tons of oil exports.

Natural Gas Importing Countries	Amount of Imported Natural Gas
China	122 billion cubic meters
Japan	105 billion cubic meters
Germany	103 billion cubic meters
Italy	71 billion cubic meters

---

[407] "Oil," last updated date: 2 February 2022, https://www.bp.com/en/global/corporate/energy-economics/statistical-review-of-world-energy/oil.html#oil-reserves

[408] Hilmar Rempel, "Geographical Distribution of Oil and Natural Gas Deposits-Different Means of Transportation to the Consumption Centers." In Pipeline Technology 2006 Conference Papers, Hannover, Germany. 2006.

[409] Metric tonne.

[410] "Statistics Report Key World Energy Statistics 2020," last updated date: 3 February 2022, https://www.iea.org/reports/key-world-energy-statistics-2020, 13.

Natural Gas Importing Countries	Amount of Imported Natural Gas
Mexico	57 billion cubic meters
South Korea	54 billion cubic meters
Turkey	44 billion cubic meters
France	44 billion cubic meters

Table 16: Top Natural Gas Importing Countries in 2018[411]

Countries do not only export oil to meet their energy needs. Many countries are also dependent on foreign sources for natural gas. China is one of these countries. China imported 122 billion cubic meters of natural gas in 2018. Japan comes right after. Japan imported 105 billion cubic meters of natural gas in 2018. Germany comes in third place. Germany, which has a smaller population than China and Japan, imported 103 billion cubic meters of natural gas in 2018. Italy, another European country, ranks fourth in the list with 71 billion cubic meters of natural gas imports.

Coal Importing Countries	Amount of Imported Coal
China	296 Mt
India	246 Mt
Japan	185 Mt
South Korea	130 Mt
Vietnam	43 Mt
Germany	41 Mt
Turkey	38 Mt
Malaysia	35 Mt

Table 17: Countries that Imported the Most Natural Gas in 2019[412]

Another of the most imported fossil fuels is coal. Especially Far Asian countries meet a significant part of their coal needs through imports. The country that imports the most coal is China. It is the largest importer of coal, as in other fossil fuels, imported 296 metric tons of coal in 2019. India is in second place. In 2019, India imported 246 metric tons of coal. Japan, which is in the third place, imported 185 metric tons, South Korea, which is in the fourth place, imported 130 metric tons, and Vietnam, which is in the fifth place, imported 43 metric tons of coal. Germany and Turkey,

411 “Statistics report Key World Energy Statistics 2020,” 15.

412 “Statistics report Key World Energy Statistics 2020,” 17.

which have similar populations, imported 41 metric tons and 48 metric tons of coal, respectively.

Electricity Importing Countries	Amount of Imported Electricity
USA	44 TWh
Italy	44 TWh
Brazil	35 TWh
Thailand	26 TWh
Iraq	22 TWh
Finland	20 TWh
The United Kingdom	19 TWh
Belgium	17 TWh

Table 18: Countries that Imported the Most Electricity in 2019[413]

Many countries also import electricity. Electricity, which is widely used in residences, industrial facilities, service sector, and transportation, is imported by countries located in different parts of the world. The USA is one of these countries. In 2018, the USA imported 44 TWh of electricity. In the same year, Italy imported 44 TWh of electricity. Brazil, which ranks third, imported 35 TWh, and Thailand, which ranks fourth, imported 26 TWh of electricity. Iraq, which has rich fossil fuel reserves, is one of the countries that import the most electricity. Iraq imported 22 TWh of electricity in 2018. Finland ranks sixth with 20 TWh electricity imports, the United Kingdom ranks seventh with 19 TWh electricity imports, and Belgium ranks eighth with 17 TWh electricity imports.

The information above shows that especially the countries with a large population and economically developed are dependent on other countries to meet their energy needs. The reason for foreign dependency in energy is that fossil fuels are an energy source that is concentrated in certain regions. Fossil fuels are not spread over a wide geography. For this reason, certain countries control fossil energy sources. As a result of this, a significant part of the world's countries must import to meet their energy needs, since they do not have fossil energy resources in their own lands.

413 "Statistics report Key World Energy Statistics 2020," 33.

The money paid to supply energy is one of the important import costs of many countries. Since fossil fuels are found in certain countries, these energy sources are obtained through imports and this situation negatively affects the economies of importing countries. For example, Turkey, which imports approximately 70% of the energy it uses, spent 30 billion dollars for energy imports in 2020.

However, uranium resources, which is the main fuel of nuclear energy, are located in wide geography compared to fossil fuels.[414] As seen in Table 19, uranium is found in different countries and geographical regions. In the list of countries producing the most uranium in the world in 2020, there are countries from different continents such as Namibia, China, India, Canada, Australia, USA, and Brazil. Some of these countries are highly dependent on foreign fossil fuels. The world's most populated countries, India and China, meet their fossil fuel needs by importing them. These are also the countries that produce the most uranium in the world. From this point of view, it should be noted that countries that are dependent on foreign fossil fuels can reduce or completely terminate energy imports by using nuclear energy, thanks to their uranium reserves.[415] For this reason, another advantage of nuclear energy is that it contributes to the reduction of foreign dependency, especially in fossil fuels.[416]

---

[414] Ch Poinssot et al., "Assessment of the Environmental Footprint of Nuclear Energy Systems. Comparison between Closed and Open Fuel Cycles," *Energy* 69 (2014): 199-211, 199.

[415] Turkey is one of the countries that is not in a geography rich in fossil fuels, but can reduce foreign dependency in energy by producing nuclear energy with its uranium resources. Located in the Middle East, which is a region rich in fossil fuels, Turkey does not have enough fossil energy resources to meet its own needs within its borders. However, Turkey has tens of thousands of tons of uranium reserves within its borders. There are more than 12 thousand tons of uranium reserves in Manisa-Köprübaşı, Uşak-Eşme-Fakılı, Aydın-Küçükçavdar, Yozgat-Sorgun and Aydın-Demirtepe regions. With the use of this reserve in the nuclear energy sector, foreign dependency on fossil fuels can be reduced or completely avoided.

[416] Nicholas Apergis and James E. Payne, "A Panel Study of Nuclear Energy Consumption and Economic Growth," *Energy Economics* 32/3 (2010): 545-549, 549.

COUNTRY	URANIUM RESERVES (TONS)
Kazakhstan	19,477
Australia	6203
Namibia	5413
Canada	3885
Uzbekistan	3500
Nigeria	2991
Russia	2846
Chinese	1885
Ukraine	744
India	400
South Africa	250
Iran	71
Pakistan	45
Brazil	15
USA	6

Table 19: Countries Producing the Most Mineral Uranium in 2020[417]

In short, the world's most populous and most developed economies such as the USA, China, and India are heavily dependent on foreign sources for fossil fuels such as oil, natural gas, and coal. Since fossil energy sources are found in certain countries, many countries meet their energy needs by importing. However, uranium, which is the main fuel of nuclear energy, is easier to access due to its wider geography. Therefore, nuclear energy has a key role in reducing foreign dependency on fossil fuels.

2.8. Low Land Use

One of the most important sectors affecting land use is energy.[418] The land is used at all stages such as raw material exploration, extraction, production, processing, distribution, and use for energy.[419] Power plants are built on the physical elements of the environment. Solar panels, wind

[417] "World Uranium Mining Production," last updated date: 3 February 2022, http://www.world-nuclear.org/information-library/nuclear-fuel-cycle/mining-of-uranium/world-uranium-mining-production.aspx

[418] GordonWalker, "Energy, Land Use and Renewables: A Changing Agenda," *Land Use Policy* 12/1 (1995): 3-6.

[419] Virginia H. Dale, Rebecca A. Efroymson and Keith L. Kline, "The Land Use," 758.

turbines, hydroelectric and thermal power plants, oil, coal and natural gas power plants, and nuclear power plants are built on land and water resources. In addition, energy sources such as oil, coal, and natural gas emerge with the transformation of the physical elements of the environment, namely mining activities. In both cases, there is a situation where the natural environment is transformed and destroyed for energy production.

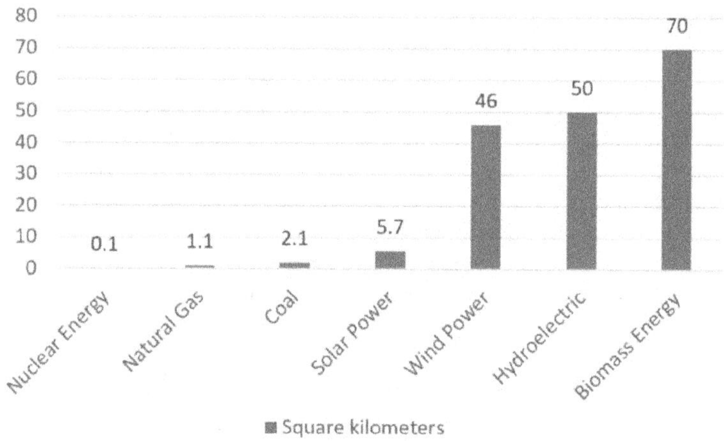

Table 20: Amount of Land Required for 1 Terawatt hour (TWs) Energy Production (km^2)[420]

The amount of land used for energy resources differs from each other. As seen in Table 20, the type of energy source that requires the most land use is renewable energy. Among the renewable energy sources, the one that requires the most land use is biomass energy. The amount of land required to produce 1 terawatt hour of energy with biomass energy is 70 km^2. The hydroelectric power station is in the second place. The amount of land required for the production of 1 terawatt hour of energy with the hydroelectric power plant is 50 km^2. With wind energy, which ranks third, the amount of land required for 1 terawatt hour of energy production is 46 km^2.

Solar energy is the energy source that requires the least land use

[420] Barry W. Brook and Corey J. A. Bradshaw, "Key Role for Nuclear Energy in Global Biodiversity Conservation," *Conservation Biology* 29/3 (2015): 702-712, 707.

among renewable energy sources.[421] While biomass energy, hydroelectric power plant, and wind energy require high land use, the amount of land required by solar energy is much less compared to other renewable energy sources. The amount of land required to produce 1 terawatt hour of energy with solar energy is 5.7 km^2. From this point of view, it is possible to say that solar energy is more advantageous than other renewable energy sources. In addition, solar panels have more advantages compared to other renewable energy sources, as they can be used on the roofs or walls of existing structures[422]. However, the lands occupied for renewable energy sources are not only used in energy production. The land is also used for mining the raw materials required for the battery of renewable energy sources.[423]

Fossil fuels are the type of energy source that requires the most land use after renewable energy sources. Mining, which is necessary to access fossil fuels, operation and storage of fossil fuels, and power plants established to generate electricity from fossil fuels require a high amount of land use. Among the fossil fuels, coal is the one that requires the most land use. The amount of land required for 1 terawatt hour of energy production with coal power plants is 2.1 km^2. Natural gas ranks second among fossil fuels. The amount of land required to produce 1 terawatt hour of energy with natural gas is 1.1 km^2. This information shows that the amount of land required for fossil fuels is less than for renewable energy sources.

Among the energy sources, the one that requires the least land use is nuclear energy. One of the important advantages of nuclear energy is the low amount of land used for energy production compared to other energy sources. The amount of land required for 1 terawatt hour of energy production in nuclear power plants is 0.1 km^2. From this point of view, it would not be wrong to state that nuclear energy is the most advantageous energy source in terms of saving land and preventing the destruction of the

---

[421] Vasilis Fthenakis and Hyung Chul Kim, "Land Use and Electricity Generation: A Life-cycle Analysis," *Renewable and Sustainable Energy Reviews* 13/6-7 (2009): 1465-1474, 1465.

[422] David C. Howard et al., "The Impact of Sustainable Energy Production on Land Use in Britain through to 2050," *Land Use Policy* 26 (2009): S284-S292, S289.

[423] Barry W. Brook and Corey J. A. Bradshaw, "Key Role for Nuclear," 710.

natural environment. However, in nuclear energy, the land is not only used for energy production. The land is also used for the storage of nuclear waste in nuclear energy and for uranium mining.

Looking at the other side of the coin, it should be noted that a large amount of land may become unusable due to nuclear accidents. This problem came to the fore with the Chernobyl and Fukushima nuclear power plant accidents. Due to both nuclear power plant accidents, a certain amount of land is unusable. For example, after the Chernobyl disaster, 30 kilometers of land in the area where the accident took place was declared as an "exclusion zone"[424] due to radioactive contamination, that is, the human settlement was prohibited. Similarly, after the Fukushima Nuclear Power Plant accident, an area of 20 kilometers was declared as an exclusion zone due to radioactive contamination in the area where the accident took place.[425] It is not known when the exclusion zones in the regions where the nuclear power plant accident occurred will be terminated. Contrary to nuclear energy, it is not possible to talk about serious problems like this in the event of an accident in renewable energy sources.

3. CONCLUSION

In this chapter, the advantages of nuclear energy in terms of environment, public health, and economy are analyzed. The results of this analysis show that nuclear energy has eight main advantages. The first is that nuclear power plants have a long operating lifetime. Energy is obtained by converting natural resources. Both fossil fuels, nuclear energy, and renewable energy sources become usable as energy as a result of the processes carried out in power plants. Each power plant has an average operating lifetime. Power plants are built with resources that are limited in nature. For this reason, power plants with a long operating lifetime are

---

[424] Ganna Yankovska and Kevin Hannam, "Dark and Toxic Tourism in the Chernobyl Exclusion Zone," *Current Issues in Tourism* 17/10 (2014): 929-939, 930.

[425] Hui Zhang et al., "Radiation-driven Migration: The Case of Minamisoma City, Fukushima, Japan, after the Fukushima Nuclear Accident," *International Journal of Environmental Research and Public Health* 11/9 (2014): 9286-9305, 9290.

needed to protect resources. The power plants with the longest operating lifetime are nuclear power and hydroelectric power plants.

The second advantage of nuclear energy is that it causes low greenhouse gas emissions. One of the most important environmental problems of the period we live in is climate change, which deeply affects the whole world. The increase in the average temperature of the earth, that is, global warming, triggers irregular changes in the climate in a short time. The main cause of greenhouse gases that cause global warming is fossil fuels. Fossil fuels, which are used to meet the constantly increasing energy needs of people after the Industrial Revolution, trigger the release of greenhouse gases such as carbon dioxide, nitrogen, and methane, causing an increase in the average temperature of the earth. For this reason, energy sources with low greenhouse gas emissions are needed to combat climate change. Nuclear energy, which is one of these energy sources, has an important advantage in the fight against climate change as it does not directly cause greenhouse gas emissions during energy production.

The third important advantage of nuclear energy is that it does not cause air pollution. One of the important environmental problems of the 21st century is air pollution. Polluted air due to human actions carries significant risks to life. Millions of people die every year due to air pollution-related diseases. In addition, other living species are also at serious risk due to the deterioration of the natural components of the air. One of the most important factors causing air pollution is the use of fossil fuels. Especially with the combustion of coal and oil, pollutants are released into the atmosphere, and this triggers air pollution. In order to prevent or reduce air pollution, clean energy sources are needed. One of these energy sources is nuclear energy. No pollutant is released during energy production in nuclear power plants. Therefore, nuclear energy can play a key role in establishing clean air.

Fourth, nuclear energy is an energy source with high fuel efficiency. Renewable energy is the energy source with the most fuel efficiency. Nuclear energy comes next. It is possible to obtain higher efficiency with uranium, which is the main fuel used in nuclear power plants, compared to fossil fuels. The amount of energy to be obtained with 1 kilogram of uranium corresponds to thousands of kilograms of fossil fuels. Therefore, nuclear energy has an important advantage in ensuring fuel efficiency. Due to the limited availability of natural resources in nature, the preference for energy sources with high fuel efficiency is important in protecting the physical and biological elements of the environment.

The fifth important advantage of nuclear energy is that it provides energy security. Energy security means having affordable and stable access to energy resources. It is not possible to get the same efficiency all the time because renewable energy sources are affected by seasonal conditions. Fossil fuels, on the other hand, are sensitive to energy security as they are affected by the volatility of prices and geopolitical developments. Unlike these two energy sources, nuclear energy is not affected by both seasonal conditions and geopolitical developments. For this reason, nuclear energy has an important advantage in ensuring energy security.

Another important advantage of nuclear energy is its contribution to the economy. One of the most important problems of modern economies is unemployment. One of the important sectors that will contribute to employment is the energy industry. One of the most important energy sectors that increase employment is nuclear energy. Hundreds of thousands of people are directly employed in the nuclear energy sector, and millions of people are indirectly employed in secondary jobs related to nuclear energy. Due to its long operating lifetime, it creates full-time and long-term employment opportunities for thousands of people. In addition, it offers employment opportunities for people from many different professions such as civil engineers, health physicists, mechanical engineers, scientists, lawyers, and heavy equipment operators.

Nuclear energy also makes a significant contribution to reducing the foreign dependency of countries, especially on fossil fuels. The energy needs of countries with growing economies and increasing populations are increasing in parallel. A significant part of the energy need is met from fossil fuels. However, fossil fuels are located in a certain geography, that is, they are under the sovereignty of certain countries. Countries such as China, Japan, and India import fossil fuels from other countries to meet their energy needs. This situation causes countries to be dependent on foreign energy and causes serious economic costs. However, since uranium, the fuel of nuclear energy, spreads to wider geography, all countries can access this resource more easily and at affordable prices. This situation can contribute positively to the reduction of foreign dependency in energy.

Finally, one of the advantages that makes nuclear power superior to other energy sources is that it requires low land use. For each energy source, the land is used and transformed at all stages such as raw material exploration, extraction, production, processing, distribution, and use. The amount of land used for energy sources differs from each other. The type of energy source that requires the most land use is renewable energy. A high percentage of land is used especially for biomass energy, hydroelectric energy, and wind energy. After renewable energy sources, fossil fuels are the type of energy that requires the most land use. The amount of land used to generate energy from coal, oil and natural gas resources is low compared to renewable energy resources. Among the energy sources, nuclear energy is the one that requires the least amount of land use. The construction of nuclear power plants in smaller areas compared to other energy sources is an important situation in terms of protecting the natural environment and biodiversity.

CHAPTER V:

DISADVANTAGES OF NUCLEAR ENERGY

1. INTRODUCTION

All energy sources have advantages as well as disadvantages in terms of environment, economy, and public health. It is not possible to talk about a perfect energy source in all aspects. Despite all technological developments, every energy production has costs in terms of the environment and public health. Looking at the other side of the coin, it is seen that nuclear energy, which has important advantages, has many critical disadvantages. It is possible to say that there are important risks arising from the use of nuclear energy, especially at the point of providing a clean environment and protecting public health. As a matter of fact, the Chernobyl Nuclear Power Plant accident that occurred in Ukraine in 1986 and the Fukushima Nuclear Power Plant accident that occurred in Japan in 2011 show us that the risks caused by nuclear energy are at an alarming level.

Nuclear power has many significant risks. Nuclear energy is not included in the energy policy of many countries due to its risks. Countries that do not use nuclear energy are relatively less populated and economically less developed. Some of these countries are as follows; Italy, New Zealand, Colombia, Peru, Australia, Nepal, and Greece. In addition, some countries are shutting down or deciding to shut down the nuclear power plants that they have been using for a long time, that is, they give up their nuclear energy policy. Germany, one of the largest economies in the world, is one of these countries. After the Fukushima Nuclear Power Plant accident, Germany decided to shut down the nuclear power plants used in the country gradually and not to build a new nuclear power plant.

The main purpose of this section is to discuss the disadvantages of nuclear energy in terms of environment, public health, and economy. The risks of nuclear energy production are examined by comparing them with other energy sources. The disadvantages of nuclear energy discussed in this section are as follows; (1) the environmental cost of nuclear energy is high, (2) nuclear energy is not renewable energy, (3) nuclear energy is not a sustainable energy source, (4) there is a risk of producing nuclear weapons from nuclear energy, (5) energy production in nuclear power plants nuclear waste, (6) nuclear energy is costly and (7) the risk of being exposed to terrorist attacks.

2. DISADVANTAGES OF NUCLEAR ENERGY

Nuclear energy has advantages as well as disadvantages. This section focuses on the most important disadvantages of nuclear energy. This section examines six main disadvantages of nuclear energy.

2.1. High Environmental Cost

The cost of nuclear energy is a matter of debate. Nuclear energy is preferred because it is one of the cheapest energy sources in electricity generation.[426] However, the cost of nuclear energy is not limited to the cost of energy production. The environmental cost of nuclear power also needs to be considered. The cost of nuclear energy is high due to its environmental effects. In addition, the construction of nuclear energy requires high costs.[427] But the environmental cost is not limited to nuclear energy. Due to their devastating effects on the ecosystem, fossil fuels also have a high environmental cost.[428]

The environmental cost of nuclear energy emerges as an important problem in the event of an accident in nuclear power plants. As mentioned earlier, one of the most important risks of nuclear power plants is the environmental problems that arise in the event of a possible accident,

[426] Allison M. Macfarlane and Marvin Miller, "Nuclear Energy and Uranium Resources," *Elements* 3/3 (2007): 185-192, 186.
[427] Richard Green and Iain Staffell, "Electricity in Europe," 287.
[428] Robert Foster, Majid Ghassemi and Alma Cota, Solar Energy, 2.

namely radioactive pollution.[429] Nuclear power plants are built and operated by taking high safety precautions. Despite all safety precautions, it is not possible to eliminate the risk of an accident. Because there is zero risk in anything. Thanks to technological progress, security measures can reduce the risk, but cannot completely eliminate it. For this reason, nuclear power plants built and operated with high safety technologies are at risk of accidents.

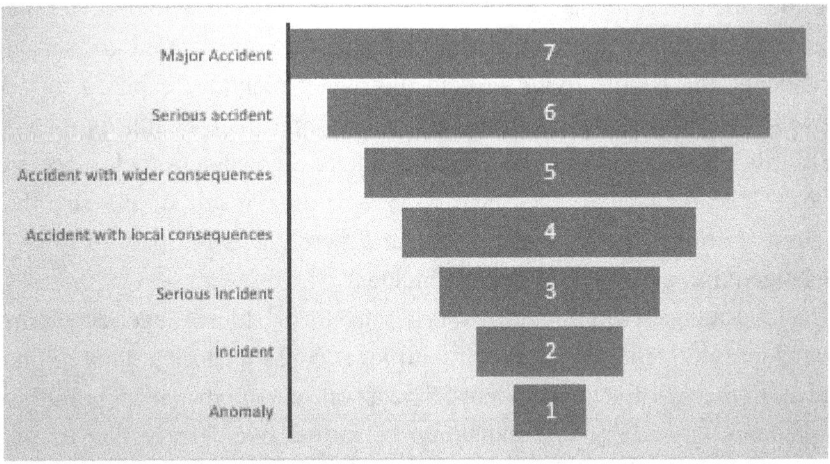

Table 21: International Nuclear and Radiological Event Scale[430]

A small number of nuclear accidents and incidents have occurred in nuclear power plants that have been used for about 70 years. Nuclear accidents and incidents are categorized by the International Nuclear and Radioactive Event Scale which shows that second-level nuclear events are grouped as incidents. In nuclear cases that fall into this group, the public is exposed to more than 10 mSv of radiation. The radiation rate in the area where the accident is located is more than 10 mSv. The radiation exposure of nuclear power plant workers is higher than the annual radiation level. In second-level nuclear incidents, there is no radioactive pollution that poses a serious risk to the environment. It usually occurs due to an error in

[429] W. Häfele, "Hypotheticality and the New Challenges: The Pathfinder Role of Nuclear Energy," *Minerva* 12/3 (1974): 303-322, 307.

[430] "International Nuclear and Radiological Event Scale (INES)," last updated date: 5 February 2022, https://www.iaea.org/sites/default/files/ines.pdf

security measures. The safety problem caused by the exposure of workers to radiation at levels exceeding the annual limit at Atucha Nuclear Power Plants in Argentina in 2005 and the failure in the emergency power supply of Forsmark Nuclear Power Plants in Sweden in 2006 are considered within the scope of the second level nuclear incident.[431]

According to the International Nuclear and Radioactive Event Scale, third-level nuclear incidents are categorized as serious incidents. In serious incidents, nuclear plant workers are exposed to radiation 10 times the annual ration limit. In accidents that are considered as serious incidents, the people living around the power plant are exposed to low levels of radiation. Radiation can cause non-fatal health problems. Security measures around the reactors where the accident took place are weak. The damage to the security system as a result of the fire that occurred around the Vandellos Nuclear Power Plants in Spain in 1989 is considered as a third-level nuclear incident.[432]

Radioactive contamination is fatal in the fourth level accidents, which are defined as an accident with local consequences at the regional level. In the fourth level accidents, at least one person dies from pollution. Accidents caused by the realization of radioactive release due to fuel melting or damage to the fuel are evaluated in this category. There is a risk of exposure of the public to high levels of radiation in accidents that fall into this group. The event that resulted in the death of hundreds of workers after the nuclear power plant accident at the Tokaimura JCO Nuclear Power Plant in 1999 is categorized within this group.[433]

Level five accidents, defined as an accident with wider consequences, are caused by damage to the reactor core. In nuclear power plant accidents that fall into this group, a high amount of radiation is released, and partial precautions must be taken. Radioactive contamination, which occurs in accidents at the fifth level, also causes human deaths. Because people living in the area where the accident occurred are exposed to high levels of radiation. Two nuclear power plant accident events are included in the accident group with broad

[431] "International Nuclear and Radiological Event Scale (INES)."
[432] "International Nuclear and Radiological Event Scale (INES)."
[433] "International Nuclear and Radiological Event Scale (INES)."

consequences. The first of these is the accident that occurred at the Windscale Nuclear Power Plant in the United Kingdom in 1957.[434] In the accident that occurred as a result of the fire in the nuclear facility, hundreds of people died due to cancer cases caused by radioactive contamination. Another nuclear power plant accident in the fifth level group is the case at the Three Mile Island Nuclear Power Plant in the USA. This accident, which occurred as a result of the damage to the reactor core, triggered a significant increase in cancer cases among the people of the region.

According to the International Nuclear and Radioactive Event Scale, sixth level nuclear events are grouped as a serious accident. In nuclear power plant accidents, which fall into this group, a high rate of radiation is released into the environment. Therefore, it is necessary to take precautions against radioactive contamination. So far, a nuclear power plant accident at the sixth level has occurred only once. The Kyshtym disaster that occurred in Russia in 1957 is considered within this group.[435] This accident, which occurred as a result of the explosion in the facility where nuclear waste is stored and processed, caused thousands of people to be exposed to radiation at a high rate in an area of 1000 kilometers.

The group that poses the highest risk to the environment and public health is the seventh level accidents, which are categorized as major accidents. Two nuclear power plant accidents are included in the major accident group. The first of these is the Chernobyl Nuclear Power Plant accident that occurred in Ukraine in 1986. Thousands of people lost their lives in this accident, which was due to human error, and died of deadly diseases such as cancer.[436] Since the radioactive contamination caused by the accident continued, the 30-kilometer area was closed to human habitation.[437] The second accident included in the major accident group is the Fukushima Nuclear Power Plant accident that occurred in Japan in 2011. The Fukushima accident, which occurred after the tsunami that occurred after the 9-magnitude earthquake damaged the reactors, caused

---

434 "International Nuclear and Radiological Event Scale (INES)."

435 "International Nuclear and Radiological Event Scale (INES)."

436 Rick J. Scheidt, "The Babushkas of Chernobyl," *The Gerontologist* 58/1 (2018): 193-195, 193.

437 Ganna Yankovska and Kevin Hannam, "Dark and Toxic Tourism," 930.

radioactive pollution in the surrounding areas that negatively affected living life.[438]

Figure 2: Radiation Measurement in Front of the Chernobyl Nuclear Power Plant [439]

The environmental disaster caused by nuclear power plant accidents has multifaceted effects. It is possible to examine these effects under three headings. First, nuclear power plants trigger environmental pollution.[440] Because accident in nuclear power plants causes radioactive pollution. Second, radioactive pollution from nuclear accident causes damage to biodiversity.[441] For example, after the Fukushima Nuclear Power Plant accident, it was determined that the size of the butterflies reduced in the Fukushima region.[442] Third, due to an accident in nuclear reactors, environmental pollution may occur that can fatally affect public health.[443]

---

[438] Carola Hommerich, "Trust and Subjective," 46.

[439] This photo was taken during my fieldwork for my doctoral study in the region where the Chernobyl Nuclear Power Plant accident took place in 2017.

[440] Jin HoSong, "An Assessment on the Environmental Contamination Caused by the Fukushima Accident," *Journal of Environmental Management* 206 (2018): 846-852.

[441] Henrik von Wehrden et al., "Consequences of Nuclear Accidents for Biodiversity and Ecosystem Services," *Conservation Letters* 5/2 (2012): 81-89.

[442] Atsuki Hiyama et al., "The Biological Impacts of the Fukushima Nuclear Accident on the Pale Grass Blue Butterfly," *Scientific Reports* 2/1 (2012): 1-11.

[443] Achilles Adamantiades and Ioannis Kessides, "Nuclear Power for Sustainable," 5159.

Figure 3: An Abandoned Residence in the Area of the Chernobyl Nuclear Power Plant Accident [444]

Radioactive contamination caused by the accident in nuclear power plants triggers deadly diseases such as cancer. According to scientific studies conducted in Fukushima and Chernobyl regions, health problems have increased after the nuclear power plant accident. One of these health problems is cancer cases. There is an increased risk of cancer among those exposed to radioactive pollution.[445] Some of these cancer cases result in death.

Figure 4: Abandoned Mass Housing in the Region where the Chernobyl Nuclear Power Plant Accident Occurred[446]

---

444 This photo was taken during my fieldwork for my doctoral study in the region where the Chernobyl Nuclear Power Plant accident took place in 2017.

445 Wolfgang Weiss, "Chernobyl Thyroid Cancer: 30 Years of Follow-up Overview," *Radiation Protection Dosimetry* 182/1 (2018): 58-61.

446 This photo was taken during my fieldwork for my doctoral study in the region where the Chernobyl Nuclear Power Plant accident took place in 2017.

On the other hand, it should be noted that although the cost of environmental pollution in the event of an accident is very high, the risk of accidents in nuclear power plants is low.[447] Nuclear power plants are built with high safety technology. Safety measures are taken at the highest level to reduce the risk of accidents. For this reason, nuclear energy accidents that pose a risk to the environment and public health are very rare.

2.2. Unsustainable Energy Source

Energy is one of the most important instruments that determine the quality of life of people in modern society. Due to the rapid increase in the world population and economic growth, the need for energy is increasing in parallel. Like the current generation, the next generation will also need energy. Because energy is consumed at every stage of human life. It is possible for future generations to access energy in a healthy way with the sustainability of energy resources. In other words, it seems unlikely to meet the energy need in a stable manner with resources that are at risk of depletion. Otherwise, people who will come to the world after us may face serious difficulties in accessing energy. Therefore, it is important that energy sources are renewable.

As stated in the previous sections, energy resources are divided into two basic groups as renewable and non-renewable. Energy sources such as solar energy, wind energy, and wave energy, which have a very low risk of depletion, are in the group of renewable energy sources.[448] These energy sources are constantly renewing themselves. For example, solar energy is obtained by converting sunlight into energy. The sun's rays fall on the earth in the process from sunrise to sunset. After the sun sets, the source of solar energy is renewed with sunrise. For this reason, the energy obtained from sources such as wind, waves, sun, and water is called renewable energy.

Unlike renewable energy sources, coal, oil, and natural gas, that is, fossil fuels, are considered as energy sources that are limited in nature

---

[447] Aviel Verbruggen, "Renewable and Nuclear Power," 4045.
[448] N. L. Panwar, S. C. Kaushik and Surendra Kothari, "Role of Renewable Energy."

and cannot be renewed in case of depletion.[449] There is no definite information about when fossil fuels will run out. Because as technology develops, fossil fuel reserves discovered change. According to studies conducted in this area, it is estimated that fossil fuels will be depleted in approximately 100 years. For example, according to a study by Shahriar Shafiee and Erkan Topal, it is estimated that oil will run out in 35 years, natural gas in 37 years, and coal in 107 years.[450] Fossil fuels are not sustainable because they are non-renewable and are a limited source of energy in nature.[451] Therefore, fossil fuels may be insufficient to meet the energy needs of future generations.

It is a matter of debate whether nuclear energy, which has emerged as an alternative to fossil fuels, will be sufficient to meet the energy needs of future generations in a sustainable way. Nuclear energy is also included in the group of non-renewable energy sources. The basic fuel required for energy production in nuclear power plants is uranium.[452], which is found in nature in limited quantities.[453] According to the World Nuclear Association, the world's uranium reserves are around 6 million tons.[454] Also, uranium is a non-renewable natural resource like fossil fuels. If uranium is depleted, it may be impossible to produce nuclear energy. For this reason, it is not possible to meet the energy needs of future generations with nuclear energy in a sustainable way.

2.3. Nuclear Weapons

Throughout human history, wars have always occurred. Wars and conflicts between human groups in both primitive and modern times have never ceased. Different tools were used to gain superiority in wars. With the advancement of information and technology over time, the destructive effect of the tools used in wars has also increased. The most destructive of

---

449 V.Manieniyan, M. Thambidurai and R. Selvakumar, "Study on Energy."

450 Shahriar Shafiee and Erkan Topal, "An Econometrics View," 181.

451 Umair Shahzad, "The Need for Renewable," 16.

452 Ayhan Demirbas, "Global Renewable Energy Projections," 215.

453 Edmund Henrich, Nicolaus Dahmen, Eckhard Dinjus, and Jörg Sauer, "The Role of Biomass," 1669.

454 "World Nuclear Association, Supply of Uranium," last updated date: 10 March 2022, https://www.world-nuclear.org / information-library / nuclear-fuel-cycle / uranium-resources/ supply-of-uranium.aspx

these tools is the nuclear weapon. After the Second World War, the countries of the world entered into a nuclear arms race together. Only in the second half of the 20th century, 128,000 nuclear weapons were produced. Of these, 70,000 were produced by the United States, 55,000 by the Soviet Union, 1100 by France, 850 by the United Kingdom, and 600 by China.[455]

Nuclear weapons were first used in the Second World War. The first use of nuclear weapons in a war was the bombing of Hiroshima, Japan, by the USA on August 6, 1945.[456] The nuclear bomb used by the USA caused the death of more than 100,000 civilians,[457] hundreds of years of radioactive contamination, destruction of the artificial environment, and damage to biodiversity. Therefore, nuclear weapons pose a great risk to humanity and the environment.

There is a close relationship between the production of nuclear weapons and nuclear energy.[458] In fact, nuclear energy emerged as a result of the nuclear armament process. However, nowadays, nuclear energy is used independently due to the restriction of nuclear armament by international law. In nuclear power plants, 1% of uranium's weight is converted to plutonium in energy production.[459] Plutonium, which is obtained artificially and is one of the most toxic elements, is used in nuclear weapons. For this reason, countries with nuclear power plants have the potential to produce nuclear weapons.

However, it should be noted that it is not possible to say that every country that uses nuclear energy or has nuclear power plants has nuclear weapons. Producing nuclear weapons from nuclear energy emerges as a result of a political attitude, that is, the choice of political power. There are many countries that have not taken this policy decision. For example, Japan, one of the most important nuclear energy-producing countries in

---

[455] Frank Barnaby, "The Effects of the Atomic Bombings of Hiroshima and Nagasaki," *Medicine, Conflict and Survival* 11/3 (1995): 1-9, 1.
[456] Michihiko Hachiya, Hiroshima Diary: The Journal of a Japanese Physician, August 6-September 30, 1945 (Chapel Hill: UNC Press Books, 2011).
[457] Frank Barnaby, "The Effects of the Atomic."
[458] D. Yogi Goswami and Frank Kreith eds., *Handbook of Energy*, 8.
[459] Allison M. Macfarlane and Marvin Miller, "Nuclear Energy," 186.

the world, does not have nuclear weapons.[460] Japan, which has lived and continues to experience the pain of the chaos caused by the nuclear weapon used by the USA for many years, has not adopted the policy of producing nuclear weapons. For this reason, there is a fine line between nuclear power generation and nuclear weapons production.

2.4. High Installation Cost

Every power plant has an installation cost. The installation costs of power plants can differ significantly from each other. Because the technological materials used in power plants, their dimensions, scopes, the number of workers employed during their construction, etc. differ from each other.[461] Nuclear power plants are power plants that are built at a high cost. Installation costs of nuclear power plants are higher than other energy sources.[462]

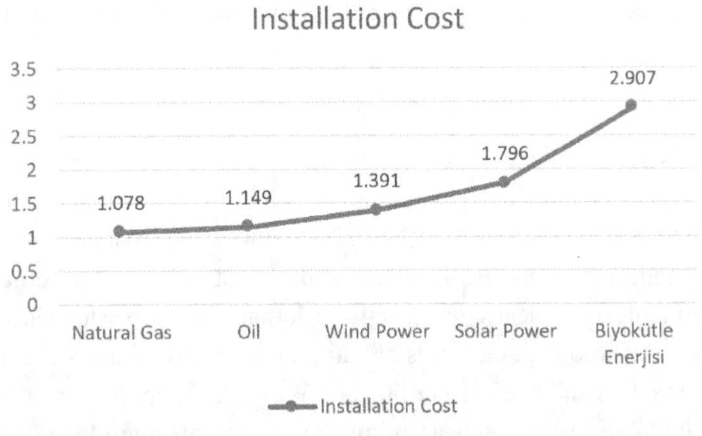

Table 22: Installation Cost of Electricity Facilities in 2019 ($/kW)[463]

As seen in Table 22, the power plants with the lowest installation costs are natural gas and oil. According to the installation cost, one

[460] Mike M. Mochizuki, "Japan Tests the Nuclear Taboo," *Nonproliferation Review* 14/2 (2007): 303-328.
[461] Lucas W. Davis, "Prospects for Nuclear Power," 53.
[462] Sabhasachi Saha et al., "Rooppur Nuclear Power Plant: Current Status & Feasibility," *Strojnícky časopis-Journal of Mechanical Engineering* 68/3 (2018): 167-182.
[463] "Construction Cost Data for Electric Generators Installed in 2019," last updated date: 19 March 2022, https://www.eia.gov/electricity/generatorcosts/

kilowatt of electricity production is 1,078 dollars in natural gas and 1,149 dollars in oil. For this reason, it is possible to say that the cost of power plants established for electricity generation from fossil fuels is lower than other energy sources. After fossil fuels, the power plants with the lowest installation cost are renewable energy. According to the installation cost, one kilowatt of electricity production is 1,391 dollars in wind power, 1,796 dollars in solar power, and 2,907 dollars in biomass energy.

The type of energy source with the highest installation cost is nuclear energy. The cost of installing nuclear power plants varies according to the reactor type. On average, one kilowatt of electricity production varies between 5,500 and 8,100 dollars in nuclear energy, according to the installation cost.[464] In other words, the establishment cost of nuclear power plants is 5-7 times that of fossil fuels and 2-6 times that of renewable energy sources. This information shows that the installation cost of nuclear power plants is quite high compared to other energy sources.

2.5. Nuclear Waste Issue

Wastes occur during energy production in every energy source. There is no production and consumption without waste. Among the energy sources, nuclear energy has the most dangerous waste in terms of environment and public health. One of the most important arguments of the opponents of nuclear energy is the risk that nuclear waste poses to the physical and biological elements of nature. Radioactive wastes occur in all nuclear reactors and nuclear facilities.[465] Wastes are generated at all stages of the fuel life cycle, such as mining, enrichment, manufacturing, and reactor operation.[466] Every year, several tons of used nuclear waste are generated worldwide.[467]

---

[464] "Nuclear Power Plant Construction Costs," last updated date: 20 March 2022, https://www.synapse-energy.com/sites/default/files/SynapsePaper.2008-07.0.Nuclear-Plant-Construction-Costs.A0022_0.pdf

[465] Walter C. Patterson, *Nuclear Power*, 11.

[466] Eugene A. Rosa et al., "Nuclear Waste: Knowledge Waste?," *Science* 329/5993 (2010): 762-763, 762.

[467] Jean-MarcMontel, "Minerals and Design of New Waste forms for Conditioning Nuclear Waste," *Comptes Rendus Geoscience* 343/2-3 (2011): 230-236.

Nuclear wastes are categorized into three main groups.[468] The first of these is high-level nuclear waste. Wastes containing high radioactive values and long-lasting are called high level nuclear wastes.[469] Used fuels in nuclear power plants fall under the scope of high level nuclear waste.[470] By 2009, 8,000 tons of high-level nuclear waste were generated in 443 nuclear power reactors around the world.[471] These wastes pose a serious risk to the environment, society, and other living species. The second is intermediate level nuclear waste. Plutonium is in the middle level nuclear waste group.[472] The third is low level nuclear waste. Reactor water treatment residues and radioactively contaminated protective overshoes and clothing are considered within this scope.

Nuclear wastes are very dangerous for the ecosystem. Some of the most dangerous chemical elements that pose a risk to the environment in nuclear waste are as follows; "*plutonium, neptunium, americium, curium, and some long-lived fission products such as iodine and technetium.*"[473] This toxicity in a nuclear waste can last for thousands of years.[474] It may take up to 100,000 years for high level and intermediate level nuclear wastes to reach a radioactive level that will not pose a risk to the environment.[475] For example, plutonium-239 takes 24,400 years to reach a radioactive level that will not pose a risk to the nature.[476] If living things

---

468 Karen Bickerstaff, ""Because We've Got History Here": Nuclear Waste, Cooperative Siting, and the Relational Geography of a Complex Issue," *Environment and Planning A* 44/11 (2012): 2611-2628, 2614.

469 T. L. Saaty and H. Gholamnezhad, "High-level Nuclear Waste Management:Analysis of Options," *Environment and Planning B: Planning and Design* 9/2 (1982): 181-196, 181.

470 James Flynn et al., "Time to Rethink Nuclear Waste Storage," *Issues in Science and Technology* 8/4 (1992): 42-48, 43.

471 Urban Strandberg and Mats Andrén, "Nuclear Waste Management in a Globalised World," *Journal of Risk Research* 12/7-8 (2009): 879-895, 882.

472 Rodney C. Ewing, Robert A. Whittleston and Bruce W.D. Yardley, "Geological Disposal of Nuclear Waste: A Primer," *Elements* 12/4 (2016): 233-237, 233.

473 Massimo Salvatores, "Nuclear Fuel Cycle Strategies Including Partitioning and Transmutation," *Nuclear Engineering and Design* 235/7 (2005): 805-816, 805.

474 C. Rodriguez et al. "Deep-Burn: Making Nuclear Waste Transmutation Practical," *Nuclear Engineering and Design* 222/2-3 (2003): 299-317.

475 Peter Hocke and Ortwin Renn, "Concerned Public and the Paralysis of Decision-making: Nuclear Waste Management Policy in Germany," *Journal of Risk Research* 12/7-8 (2009): 921-940, 921.

476 James Flynn et al., "Time to Rethink," 243.

come into contact with these wastes, fatal results may occur. For this reason, nuclear wastes must be safely stored in an environment inaccessible to living things.

Until now, no country has been able to find a permanent solution for the storage of nuclear wastes containing high radiation.[477] There are several alternatives for the storage of nuclear waste. These consist of views such as geological disposal, burial under the sea, burial under glaciers at the poles, and throwing waste into space[478]. One of the most common methods used in the storage of nuclear waste is geological disposal. High-level nuclear wastes and intermediate level nuclear wastes are preserved by the geological disposal method.[479] In this method, nuclear wastes are buried underground at a depth that people cannot reach. In the geological disposal method, the wastes are stored in steel boxes at a depth of a few hundred meters to a few kilometers from the land surface.[480]

There are two main problems with the storage of nuclear waste. First, it seems possible to ensure that the stored waste remains safely over the long term. The protection mechanism of the wastes that will wait underground for thousands of years may become dysfunctional due to natural reasons and this may trigger an environmental disaster. Because everything has its share of wear and tear. Second, environmental disasters can occur in the event of terrorist attacks on nuclear waste facilities. There is a risk that terrorist groups will attack nuclear waste facilities in order to create fear in the society by damaging the physical and biological elements of the environment and to make concessions from the political power.

2.6. Terror

The term terrorism was first used in the French Revolution in the

---

[477] Allison M. Macfarlane and Marvin Miller, "Nuclear Energy,"186.

[478] Hyungjin Kim, Chul Park and Oh Joon Kwon, "Conceptual Design of the Space Disposal System for the Highly Radioactive Component of the Nuclear Waste," *Energy* 115 (2016): 155-168.

[479] Rodney C. Ewing, Robert A. Whittleston and Bruce W.D. Yardley, "Geological Disposal," 233.

[480] W. Russell Alexander and Linda McKinley, eds., *Deep Geological Disposal of Radioactive Waste* (Oxford: Elsevier, 2011), 2.

18th century.[481] Etymologically, the concept of "terrorism" derives from the word "terror."[482] Terrorism has many different definitions. Terrorism is most commonly defined as the use of violence or the threat of violence against society, institutions, and the state in order to achieve political, religious, and ideological goals that are targeted by a group of people.[483] Terrorists targeting a certain part or all of the society are organized systematically within a certain organization.[484] Terrorism carries risks that can deeply affect the whole world.[485]

Since the second half of the 20th century, the scope of terrorism has expanded and evolved.[486] Because terrorist organizations have targeted not only civilians and public officials but also the physical and biological elements of the environment.[487] The environment has been the target chosen by terrorist groups to create fear in society, with the increase in environmental awareness all over the world since the 1960s. Terrorist groups have begun to put pressure on the legitimate power by creating fear in society by damaging the physical and biological elements of the environment in order to realize their ideological goals. While the main target in classical terrorist attacks is individuals, public employees, and government institutions, in environmental terrorism, elements of the natural environment have been targeted by terrorist organizations. Environmental terrorism is increasing its violence all over the world.[488]

---

[481] Jeffrey L. Arnold et al., "A Proposed Universal Medical and Public Health Definition of Terrorism," *Prehospital and Disaster Medicine* 18/2 (2003): 47-52, 48.
[482] Igor Primoratz, "What is Terrorism?," *Journal of Applied Philosophy* 7/2 (1990): 129-138.
[483] Terje Aven and Seth Guikema, "On the Concept and Definition of Terrorism Risk," *Risk Analysis* 35/12 (2015): 2162-2171, 2162.
[484] George P. Fletcher, "The Indefinable Concept of Terrorism," *Journal of International Criminal Justice* 4/5 (2006): 894-911.
[485] Gregor Bruce, "Definition of Terrorism Social and Political Effects," *Journal of Military and Veterans Health* 21/2 (2013): 26-30.
[486] Verena Erlenbusch, "How (not) to Study Terrorism," *Critical Review of International Social and Political Philosophy* 17/4 (2014): 470-491, 473.
[487] Elizabeth L. Chalecki, "A New Vigilance: Identifying and Reducing the Risks of Environmental Terrorism," *Global Environmental Politics* 2/1 (2002): 46-64.
[488] Paola Andrea Spadaro, "Climate Change, Environmental Terrorism, Eco-terrorism and Emerging Threats," *Journal of Strategic Security* 13/4 (2020): 58-80.

In environmental terrorism, the targets chosen by terrorist groups for the attack can be evaluated in a wide range. One of the most commonly attacked targets is forests.[489] By burning forests, it is aimed to create fear in the society and to try to make concessions from the legitimate power.[490] Another target group of illegal groups in environmental terrorism is energy resources.[491] Terrorist attacks, especially on natural gas and oil pipelines, occur at regular intervals. It is aimed by the illegal group to put pressure on political power by damaging energy facilities, energy systems, energy infrastructure, and power plants.[492]

Nuclear power plants are also at risk of attack by terrorist organizations or illegal groups. So far, there has been no successful direct attack on nuclear power plants by terrorist groups. But that doesn't mean it won't happen until now. With the terrorist attack on the Twin Towers in the USA in 2001, the concern that illegal groups could also attack nuclear power plants increased.[493] It is one of the possible situations where terrorist groups that kill people in order to get concessions from the states try to create chaos by targeting nuclear power plants. In the case of terrorist attacks on nuclear power plants, which are protected by high security measures, a much more dangerous environmental disaster can be seen than the environmental disaster in Chernobyl and Fukushima. This situation can cause destruction for both humanity and the ecosystem.

3. CONCLUSION

In this section, six main disadvantages of nuclear energy in terms of environment, public health, and economy are examined. The first disadvantage of nuclear energy is its high environmental cost. Nuclear

---

[489] Col. János Besenyő, "Inferno Terror: Forest Fires as the New form of Terrorism," *Terrorism and Political Violence* 31/6 (2019): 1229-1241.

[490] Emrah Akyüz, "Türkiye'de Çevresel Terör: PKK Terör Örgütü Örneği," *Ulusal Çevre Bilimleri Araştırma Dergisi* 34 (2020): 197-207.

[491] Shannon O'Lear, "Environmental Terrorism: A Critique," *Geopolitics* 8/3 (2003): 127-150, 141.

[492] Mehmet Efe Biresselioglu and Isik Ozge Yumurtaci, "Evaluating the Nature of Terrorist Attacks on the Energy İnfrastructure: The Periodical Study for 1970-2011," *International Journal of Oil, Gas and Coal Technology* 10/3 (2015): 325-341.

[493] Allison M. Macfarlane and Marvin Miller, "Nuclear Energy," 187.

power plants do not cause air, water, and soil pollution while generating energy. In addition, greenhouse gas emissions do not occur in nuclear power plants. Therefore, nuclear energy is considered a clean energy source. However, in the event of an accident in nuclear power plants, radioactive contamination may occur. This situation may cause environmental problems that will affect all living species to a deadly extent. The environmental problems caused by the Chernobyl and Fukushima nuclear power plant accidents and their effects on public health are examples of these accidents. For this reason, nuclear energy is accepted as an energy source with high environmental costs.

The second main disadvantage of nuclear energy is that it is not a sustainable energy source. The main fuel used for power generation in nuclear power plants is uranium which is found in nature in limited quantities. It cannot be renewed if it is used up. For this reason, it is not possible to meet the energy needs of future generations with nuclear energy in a sustainable way. In order to meet the energy needs of future generations, there is a need for self-renewable sources such as wind energy and solar energy that do not have the risk of exhaustion.

The third main disadvantage of nuclear energy is the risk of producing nuclear weapons. Throughout history, people have constantly fought each other. With the development of weapon technology, the destructiveness of wars has also increased. Especially with the use of nuclear weapons, a process that threatens all humanity and the ecosystem has begun. In order to prevent the destructive effects of nuclear weapons, the production of nuclear weapons should be restricted or prohibited. There is a close relationship between the production of nuclear weapons and nuclear energy. In nuclear power plants, 1% of its weight is converted to plutonium in energy production from uranium, which is the main fuel. Plutonium is used to make nuclear weapons. For this reason, the capacity of nuclear energy producing countries to produce nuclear weapons is increasing.

Another disadvantage of nuclear energy is the high cost of installation. Every power plant has an institution cost. The amount of this cost varies according to the energy source. Fossil fuels are the energy source with the lowest institutional cost. According to the installation cost, one kilowatt of electricity production is 1,078 dollars in natural gas and 1,149 dollars in oil. After fossil fuels, the energy source with the lowest installation cost is renewable energy. According to the installation cost, one kilowatt of electricity production is 1,391 dollars in wind power, 1,796 dollars in solar power, and 2,907 dollars in biomass energy. The energy source with the highest institutional cost is nuclear energy. According to the installation cost, one kilowatt of electricity production varies between 5,500 and 8,100 dollars in nuclear energy. For this reason, it is possible to say that nuclear energy is an expensive type of energy compared to other energy sources.

Fifth, radioactive wastes, which pose a high risk for the environment and public health, arise in nuclear power plants. Every production has an environmental cost. Because wastes arise as a result of every production. Among the energy sources, nuclear energy is the most dangerous waste in terms of environment and public health. During energy production in nuclear power plants, high level, intermediate level, and low level nuclear wastes are generated. If high level nuclear waste is not stored safely, devastating problems can arise for both the environment and society.

Finally, there is the risk of a terrorist attack on nuclear power plants. Terrorism is one of the most important problems of the period we live in. One of the target groups of terrorist groups that kill people in order to create fear in society is the physical and biological elements of the environment. Recently, terrorist groups trying to make concessions from the political power by damaging the environment have also been organizing attacks on energy resources. Terrorist groups frequently attack oil and natural gas pipelines. One of the energy sources that terrorist groups are likely to attack is nuclear energy. The environmental disaster that may occur after terrorist attacks on nuclear power plants carries a great risk for all living things.

CHAPTER VI:

PERCEPTION AND REALITY IN NUCLEAR ENERGY

1. INTRODUCTION

Nuclear energy has both advantages and disadvantages, as outlined in the previous sections. Nuclear energy policy in a country or region is formed by evaluating these advantages and disadvantages. Power generation from nuclear power plants may become a priority if the benefits outweigh. If the disadvantages are greater, the implementation of nuclear energy policy may become more difficult. Because the determination of a policy is shaped by the evaluation of risks and opportunities. Therefore, the key question to be discussed in the formulation of nuclear energy policy is as follows; "Do the advantages or disadvantages of nuclear power outweigh?"

It is not possible to give a universally accepted answer to this question. Whether the advantages or disadvantages of nuclear energy outweigh may differ from person to person. Risks and opportunities can be perceived differently by people. There are variables that affect and shape the perceptions of individuals. One of these variables is the value judgments of the individual. Everyone's value judgments may differ more or less from each other. Because our value judgments are shaped according to our education, what we have learned in our family and friends circle, what we read, watch, hear, in other words, our past experiences. Since the experiences of each individual are different from each other, their value judgments also differ. Individuals with different value judgments, on the other hand, can look at the risks and opportunities related to an event or phenomenon from different perspectives.

The main purpose of this chapter is to examine what factors shape our perception of the advantages and disadvantages of nuclear energy. The factors that determine individuals' perceptions of nuclear energy are discussed. This section consists of three main parts. In the first part, the concept of risk, the scope of risk, and the perception of risk and benefit are examined. The next section discusses the relationship between nuclear energy and risk perception. In the last part, policy recommendations are discussed.

2. RISK AND BENEFIT PERCEPTIONS

The concept of risk is frequently used in our daily life. Risk essentially means probability. Risk is the expectation that an event will happen. In other words, the risk is uncertainty about the consequences of an event. The probability of an event occurring due to natural events or human activities is called risk.[494] For example, there is a risk of an accident that may occur when a person driving a car has a heart attack and loses control of the steering wheel. Similarly, there is a risk of the collapse of the houses we live in due to a high-intensity earthquake. Within the framework of these examples, it should be noted that risk is the possible negative effects or consequences of an expected event, both in natural events and human activities. For this reason, the risk is also defined as the expected danger.[495] From this point of view, risk has negative meanings.

Everything has a risk. There are risks at every stage of our life.[496] There is no zero risk in the world. There is a risk of accident in the vehicle we are on, the risk of collapse of the building we are in, the risk of breaking down the computer, the risk of making the air we breathe sick, and the risk of extinction due to climate change. There is nothing without risk. With the development of technology in the last 200 years, the size of the risk has also increased. The risks caused by technology have begun to be seen

[494] Lennart Sjöberg "Risk Perception: Experts and the Public," *European Psychologist* 3/1 (1998): 1-12.

[495] Ortwin Renn, "The Role of Risk Perception for Risk Management," *Reliability Engineering & System Safety* 59/1 (1998): 49-62.

[496] Ignacio Rodríguez-Garzón et al., "Voluntary and Involuntary Risk Acceptance: A Case Study of Firefighters," *Safety Science* 142 (2021): 105394.

in all phases of our lives. Because technology is used at every stage of our lives. For this reason, Ulrich Beck also considers the period we live in as a "*risk society*."[497]

Risk, which is a part of our lives, can have different meanings for different people.[498] There are many factors that affect how risky we find something to be. An individual's education, gender,[499] age, social status, purchasing power, life experiences, etc. variables directly affect the risk perception. For example, according to a study by Michael W. Slimak and Thomas Dietz, it has been determined that people with higher education and economic status are less worried about environmental risks, that is, they have a lower risk perception. According to this study, the economic and educational status of the person affects the risk perception.[500] In another study, it was determined that the risk perception of different genders was also different. According to the study by Anna Olofsson and Susanna Öhman, men perceive known risks less than women.[501]

The risk and benefit perceptions are subjective judgement.[502] Risk perception is based on an individual's belief[503] and/or feelings[504] about the dangers he or she will face about an event. The degree of this belief can vary from person to person. For example, while paragliding may be perceived as a very risky activity for some, it is done regularly by others

---

[497] Ulrich Beck, *Ulrich Beck: Pioneer in Cosmopolitan Sociology and Risk Society*, (Cham: Springer International Publishing, 2014).

[498] Paul Slovic, Baruch Fischhoff and Sarah Lichtenstein, "Why Study Risk Perception?," *Risk Analysis* 2/2 (1982): 83-93, 85.

[499] Qi Bian et al., "Risk Perceptions of Nuclear Energy, Climate Change, and Earthquake: How are They Correlated and Differentiated by Ideologies?," *Climate Risk Management* 32 (2021): 100297.

[500] Michael W. Slimak and Thomas Dietz, "Personal Values, Beliefs, and Ecological Risk Perception," *Risk Analysis* 26/6 (2006): 1689-1705.

[501] Anna Olofsson and Susanna Öhman, "Vulnerability, Values and Heterogeneity: One Step Further to Understand Risk Perception and Behaviour," *Journal of Risk Research* 18/1 (2015): 2-20.

[502] Ge Xu et al., "Environmental Risk Perception and Its Influence on Well-being," *Chinese Management Studies* 11/1 (2017): 35-50

[503] Shanyong Wang et al., "Public Perceptions and Acceptance of Nuclear Energy in China: The Role of Public Knowledge, Perceived Benefit, Perceived Risk and Public Engagement," *Energy Policy* 126 (2019): 352-360.

[504] Harry Otway and Kerry Thomas, "Reflections on Risk Perception and Policy 1, 2," *Risk Analysis* 2/2 (1982): 69-82.

because it is in the low-risk group. Benefit perception is an individual's belief in the possible positive consequences that an event might have.[505] For example, the idea that nuclear power plants will meet the country's energy needs and the economic return of this can be given as an example of the perception of benefit.

On the other hand, perception and reality may differ from each other. How we perceive something does not mean that it is generally true. Individual perceptions of people and scientific facts can contradict each other. For example, our perception of someone who drives a Ferrari brand car in many countries is that he is very rich. Because Ferrari brand cars are vehicles that can be used by people with high purchasing power. However, this car of the person we perceive as rich may belong to a friend or other acquaintance, in other words, it may be someone who drives it temporarily. While the perception is that this person looks rich, the reality is that the vehicle used by this person is borrowed. Our perception is our reality until we learn otherwise.

The risk and benefit perception can vary from individual to individual as well as from society to society. Information about risks and benefits is part of cultural heritage. The knowledge adopted by society about risk is passed on to future generations.[506] For example, in many countries, it is believed that seeing a black cat will bring bad luck and this will have negative consequences such as the risk of an accident. This belief has been transferred from society to society for centuries. However, there is no scientific evidence for the negative effects of seeing a black cat. Similarly, according to the information transferred from society to society, it is believed that if bird droppings fall on a person, it will bring luck, that is, it will be beneficial. However, there is no scientific study to support this belief. Our perception of risk and benefit is also shaped by the unproven information that society imposes on us.

The risk and benefit perceptions are not immutable facts. Individuals' risk and benefit perceptions can change over time. This

---

[505] Shanyong Wang et al., "Public Perceptions and Acceptance."
[506] Anders A F Wahlberg and Lennart Sjoberg, "Risk Perception and the Media," *Journal of Risk Research* 3/1 (2000): 31-50.

change usually takes place slowly and over a long period of time.[507] The reasons for the change may be different. The risk and benefit perceptions may also change over time, depending on reasons such as obtaining new information about an event and the emergence of new experiences. For example, many studies show that after the Fukushima Nuclear Power Plant accident, the individuals' risk perception of nuclear energy has changed significantly.[508, 509] While the individuals' risk perception increased after the Fukushima disaster, the opposition to nuclear energy tended to increase as a result.

Just as the risk perception differs from person to person, the tolerability of risk also differs from person to person. How tolerable the risk is based on the individual's comparison of risks and benefits. For example, the negative effects of junk food on human health are widely known. People who regularly eat junk food tolerate the health risk posed by these foods. Because the pleasure or happiness from these foods, that is, the benefit outweighs the risks. On the other hand, those who do not tolerate this risk avoid consuming junk food when the risk outweighs the opportunity.

3. NUCLEAR ENERGY AND PERCEPTION

Nuclear energy is an energy source with both benefits and risks. As discussed in the third and fourth chapters, nuclear energy has many advantages and disadvantages in terms of society, environment, and economy. It seems possible to decide whether nuclear energy is necessary or not by weighing the risks and benefits of nuclear energy. However, as stated in the previous section, the benefits and risks of a subject can be perceived differently by different people. For example, according to a study, it was determined that scientists in universities tend to see the risks related to nuclear energy and nuclear waste as greater than scientists

---

[507] Otway, Harry and Kerry Thomas, "Reflections on Risk Perception."

[508] Lei Huang et al., "The Changing Risk Perception Towards Nuclear Power in China after the Fukushima Nuclear Accident in Japan," *Energy Policy* 120 (2018): 294-301.

[509] Gabriele Prati and Bruna Zani, "The Effect of the Fukushima Nuclear Accident on Risk Perception, Antinuclear Behavioral İntentions, Attitude, Trust, Environmental Beliefs, and Values," *Environment and Behavior* 45/6 (2013): 782-798.

working for private research laboratories.[510] Therefore, whether benefits or risks outweigh nuclear energy is directly related to how the individual perceives opportunities and risks.

There are three important methods that determine the decision-making process of individuals about nuclear energy policy.[511] The first of these is cost-benefit calculations. According to this approach, individuals can decide by comparing the advantages and disadvantages of nuclear energy. Those who think that nuclear energy brings more risk than an opportunity to the society and environment they live in may be inclined to oppose nuclear energy policy. Similarly, those who think that nuclear energy brings more opportunity than the risk to the society and environment in which they live may be inclined to support nuclear energy policy. For example, according to a study conducted by Silje Kristiansen et al., it was determined that the most important factor determining the attitudes of individuals towards nuclear energy after the Fukushima accident was the comparison of opportunities and risks.[512] In short, attitudes towards nuclear energy can be formed by making an evaluation and comparison within the framework of the knowledge they have about nuclear energy.

The second method that individuals use when making decisions in nuclear energy policy is the reliance on reference groups.[513] Individuals can make decisions within the framework of the discourses and preferences of the groups they trust and feel close to their values in matters concerning themselves and society. Political, religious, and sports groups may be the ones that an individual trusts and influence his decision. One of the most common among these is political groups. For example, the nuclear energy policies of the party that a person votes for or supports in

---

510 Richard P. Barke and Hank C. Jenkins-Smith, “Politics and Scientific Expertise: Scientists, Risk Perception, and Nuclear Waste Policy,” *Risk Analysis* 13/4 (1993): 425-439, 425.

511 James H. Kuklinski, Daniel S. Metlay and W. D. Kay. “Citizen Knowledge and Choices on the Complex Issue of Nuclear Energy,” *American Journal of Political Science* 26/4 (1982): 615-642, 617-619.

512 Silje Kristiansen et al., “Risk Perception of Nuclear Energy after Fukushima: Stability and Change in Public Opinion in Switzerland,” *International Journal of Public Opinion Research* 30/1 (2018): 24-50.

513 James H. Kuklinski, Daniel S. Metlay and W. D. Kay. “Citizen Knowledge,” 618-619.

each election may directly or indirectly affect his/her choices about nuclear energy. Attitudes towards nuclear energy may change positively in parallel with the trust in reference groups. As a matter of fact, a study by Stephen C. Whitfield et al. shows that with the increase in trust, the opposition to nuclear energy decreases.[514]

Third, individuals can make decisions about nuclear energy with their own value judgments.[515] People can decide what is most important to their lives, or whether the thing is important to their lives, based on their own feelings, thoughts, experiences, and priorities. The acceptance/ approval or rejection of the use of nuclear energy may be a decision that the individual will make with his/her own values. But it is not possible to talk about a single value judgment. Since the values that individuals have will differ partially or completely from each other, the acceptability of nuclear energy in society may also differ from person to person. In other words, our values are one of the important factors that determine how we evaluate the advantages and disadvantages of nuclear energy.

There are many scientific studies on the relationship between values and nuclear energy. People with different values also have different perspectives on nuclear energy. For example, according to a study conducted by Judith I. M. de Groot et al.,[516] it was determined that people with egoistic values and people with altruistic values differ in their risk and benefit perceptions regarding nuclear energy. People with egoistic values have the idea that there are more opportunities and fewer risks associated with nuclear energy. Accordingly, people with egoistic values are more in favor of using nuclear energy. Whereas people with high altruistic values perceive more opportunities and fewer risks related to nuclear energy. People with high altruistic values are less in favor of nuclear energy or more against it. People with high biospheric values, on the other hand, stay away from the use of nuclear energy because they do not view the benefits of nuclear energy positively.

---

[514] Stephen C. Whitfield et al. "The Future of Nuclear Power: Value Orientations and Risk Perception," *Risk Analysis: An International Journal* 29/3 (2009): 425-437.
[515] James H. Kuklinski, Daniel S. Metlay and W. D. Kay. "Citizen Knowledge," 619-620.
[516] De Groot et al., "Values, Perceived Risks and Benefits, and Acceptability of Nuclear Energy," *Risk Analysis: An International Journal* 33/2 (2013): 307-317.

People with traditional values, on the other hand, show a weaker value towards the environment and environmental problems. Because people with traditional values trust the decisions and risks taken by institutions more.[517] For example, according to a study conducted by Sara K. Yeo et al., it was observed that the perception of nuclear risk decreased among conservatives in the USA after the Fukushima disaster.[518] According to a study by Stephen C. Whitfield et al., it has been determined that people with traditional values support nuclear energy more. It has been found that people with high altruistic values are more opposed to nuclear energy.[519] There are also studies contradicting this hypothesis. For example, according to a study conducted by Qi Bian et al., in Taiwan, it was determined that people who support the conservative party have a higher risk perception towards nuclear energy.[520]

In addition to values, another factor affecting individuals' attitudes towards nuclear energy is their beliefs. The belief we have about an event or phenomenon can directly affect our perspective and attitudes towards that event or phenomenon. There are many factors that shape people's beliefs.[521] The first of these is the information obtained through the direct observations and experiences of people. For example, someone who lived in Chernobyl in 1869 and was exposed to the effects of the Chernobyl nuclear power plant accident is more likely to think that nuclear power plants are too dangerous. As a matter of fact, after the Three Mile Island accident in 1979, the Chernobyl disaster in 1986, and the Fukushima Nuclear Power Plant accident in 2011, social support for nuclear energy began to decrease significantly and the environmental effects of nuclear energy began to be questioned more.[522]

---

[517] Stephen C. Whitfield et al. "The Future of Nuclear Power," 427.

[518] Sara K. Yeo et al., "Partisan Amplification of Risk: American Perceptions of Nuclear Energy Risk in the Wake of the Fukushima Daiichi Disaster," *Energy Policy* 67 (2014): 727-736.

[519] Stephen C. Whitfield et al. "The Future of Nuclear Power."

[520] Qi Bian et al., "Risk Perceptions of Nuclear Energy."

[521] Harry J. Otway, Dagmar Maurer and Kerry Thomas, "Nuclear Power: The Question of Public Acceptance," *Futures* 10/2 (1978): 109-118, 110.

[522] Miranda A. Schreurs, "The Ethics of Nuclear Energy: Germany's Energy Politics after Fukushima," *The Journal of Social Science* 77 (2014): 9-29, 10.

However, someone who has not experienced a nuclear power plant accident before may have a very different view and attitude towards nuclear energy compared to someone who has experienced a nuclear power plant accident. For example, an individual living in the vicinity of the Dampierre Nuclear Power Plant, which has been in use since 1980 in France and has never been exposed to a nuclear power plant accident, may be more inclined to believe that nuclear power plants are safe. The change in risk perception due to power plant accidents increases the opposition to nuclear energy. For instance, according to a study conducted by Lei Huang, it was determined that opposition to nuclear energy increased in China after the Fukushima nuclear power plant accident in 2011.[523] In short, what we experience and observe is one of the most important factors that directly affect our risk perception and attitude towards nuclear energy.

Another factor affecting the attitudes of individuals towards nuclear energy is the information they have obtained outside of their own experience, that is, the information they receive from third parties.[524] This information can be objective information that can be proven scientifically, or it can also be information that has no validity and consistency, that is, it has subjective value. For example, the environmental friendliness of electric vehicles is accepted by many people. Because that is how it is presented in the media. However, from a scientific point of view, electric vehicles cause high greenhouse gas emissions, triggering air pollution and climate change.

There is also a lot of information on nuclear energy that has no scientific value and is inconsistent but is believed to be true by society. For example, it is believed by a group that individuals living around nuclear power plants are exposed to high levels of radiation. However, the radiation rate that people living around nuclear power plants are exposed to is less than the radiation we get when we eat a banana. For this reason, it is possible to say that the fact that we have scientifically unproven information about nuclear energy may negatively affect our attitude towards nuclear energy.

---

[523] Lei Huang et al., "The Changing Risk Perception."

[524] Harry J. Otway, Dagmar Maurer and Kerry Thomas, "Nuclear Power," 110.

4. POLICY RECOMMENDATIONS

The results of this study show that nuclear energy is an energy source with both advantages and disadvantages. Nuclear energy has significant opportunities as well as risks for the environment, economy, and society. Individuals' attitudes towards nuclear energy are determined by comparing the perceived risks and benefits.[525] Those with higher benefit perception than risk perception support nuclear energy more, while those with higher risk perception than benefit perception are more opposed to nuclear energy. However, as discussed before, there are many factors that affect the risk and opportunity perception of individuals. Since these factors vary from person to person, it does not seem possible to say whether the pros or cons of nuclear energy outweigh.

Five basic policy recommendations are presented by this study for the successful implementation of energy policy on nuclear energy. First, public institutions need to implement more educational activities on nuclear energy. Courses on nuclear energy and other energy sources need to be introduced in both secondary and higher education. Individuals need to know the advantages and disadvantages of nuclear energy in order to make rational decisions about nuclear energy. The more individuals know about nuclear energy, the more they support nuclear energy.[526] Increasing the level of knowledge about nuclear energy enables nuclear energy policy to be supported more.[527] However, since the risk perception of individuals differs from each other, it would not be correct to generalize on this issue.

Second, the public should be involved in the decision-making process regarding nuclear energy. Since each individual's risk and benefit perceptions are different from each other, it is important that nuclear energy decisions are taken in a way that reflects the common will of groups with different risk and benefit perceptions. One of the most important factors affecting our risk perception is whether the risk is a voluntary risk or involuntary risk. While individuals accept high risks in the decisions they make of their own will, they are less likely to tolerate risks in events

---

[525] Shanyong Wang et al., "Public Perceptions and Acceptance."
[526] Man-Sung Yim and Petr A. Vaganov, "Effects of Education on Nuclear Risk Perception and Attitude: Theory," *Progress in Nuclear Energy* 42/2 (2003): 221-235, 221.
[527] Shanyong Wang et al., "Public Perceptions and Acceptance," 358.

that occur outside their own will.[528] For example, although smoking causes fatal diseases, individuals voluntarily accept the high risks of smoking because they decide to smoke voluntarily. However, the use of nuclear energy, which is not as dangerous as cigarettes, can be seen as more dangerous because it is not a decision taken of their own will. For this reason, it is likely that the energy policy will be more adopted by society, as the risks and opportunities of nuclear energy will be voluntarily accepted or rejected if public participation is ensured in decisions about nuclear energy.

Third, the advantages and disadvantages of nuclear energy should be evaluated by taking into account variables such as the economy, geographical structure, and natural resources of the countries. It is not possible to generalize about nuclear energy policy on a universal scale. Considering a country's owning nuclear power plants as a rational policy does not result in the rational use of nuclear energy by other country or countries. Because each country has natural resources, energy needs, geographical conditions, etc. differ from each other. For example, Maldives, which is on a land of approximately 300 km^2, does not need nuclear energy as it can meet all its energy needs with its rich solar energy potential. However, since Russia, one of the world's largest economies and one of the most populated countries, is not in geography rich in renewable energy resources, one of the most effective energy sources that can be used against the environmental effects of fossil fuels is nuclear energy. For this reason, nuclear energy should be determined not according to universal values, but according to the conditions of the region.

Fourth, it is necessary to provide access to information on energy policies.[529] Access to information is a basic human right. It is a fundamental freedom for individuals to have information on issues that will affect their own lives and to access information for this purpose. All technical information about nuclear energy, the opportunities offered by

---

[528] Jürgen Rehm, Dirk W. Lachenmeier and Robin Room, "Why Does Society Accept a Higher Risk for Alcohol than for Other Voluntary or Involuntary Risks?," *Bmc Medicine* 12/1 (2014): 1-6.

[529] Toby Mendel, "Freedom of Information as an Internationally Protected Human Right," *Comparative Media Law Journal* 1/1 (2003): 39-70.

nuclear energy, and the risks it poses should be accessible at a level that everyone can understand. Individuals can determine their energy preferences by knowing more about nuclear energy through access to information. In order for nuclear energy to be accepted or rejected by society, first of all, access to information about nuclear energy at the societal level must be established by public institutions and relevant private companies. For this reason, all information on nuclear energy should be written, verbal and visual and should be accessible to everyone.

Fifth and lastly, nuclear energy policy should be designed with the inclusion of regional countries in the decision-making process. Because the possible environmental effects of nuclear power plants can significantly affect the neighboring countries. For example, after the Chernobyl Nuclear Power Plant accident that occurred in 1986, Belarus, neighboring Ukraine, was exposed to radioactive pollution, and subsequently the physical and biological elements of the environment were damaged.[530] Nuclear energy directly concerns not only the countries where the power plants will be built, but also neighboring countries. Therefore, neighboring countries should be included in the decision-making process, especially for nuclear power plants planned to be built in border regions.

5. CONCLUSION

In this section, the factors affecting individuals' decision-making about nuclear energy are examined and policy recommendations are presented within this framework. First, the risk and benefit perceptions are discussed. Risk is the probability that the negative consequences of an event will occur. Risk is based on the subjective assessment of individuals. A person's beliefs and feelings about the negative consequences of an event form his/her risk perception. Benefit perception is individuals' beliefs and feelings about the possible positive consequences of an event. Everyone's risk and benefit perceptions may differ from each other. A

---

[530] Kirsten B.Moysich, Ravi J. Menezes and Arthur M. Michalek, "Chernobyl-related Ionising Radiation Exposure and Cancer Risk: An Epidemiological Review," *The Lancet Oncology* 3/5 (2002): 269-279.

person's values, beliefs, experiences, education level, social status, gender, etc. variables affect the risk and benefit perceptions.

Secondly, the risk and benefit perceptions in nuclear energy are examined. One of the most important steps that determines an individual's attitude towards nuclear energy is to take a risk and opportunity assessment. As examined in the fourth and fifth chapters of this book, the opportunities, and risks of nuclear energy are evident in the scientific framework. However, since individuals' risk and benefit perceptions differ from each other, the advantages and disadvantages of nuclear energy are evaluated differently. Therefore, the attitudes of individuals towards nuclear energy also differ from each other. People's attitudes towards nuclear energy are not based on facts about nuclear energy's opportunities and risks, but on how these realities are perceived.

Finally, policy recommendations were presented within the framework of the analysis. It is recommended to take five basic steps in the formation of an energy policy on nuclear energy. These are as follows; (1) creating educational activities related to nuclear energy, (2) ensuring the participation of the public in the decision-making process related to nuclear energy, (3) establishing the energy policy related to nuclear energy depending on variables such as the geographical structure, economy and natural resources of the countries, (4) public access to all information on nuclear energy should be ensured; and (5) regional countries should be included in the decision-making process regarding nuclear energy.

BIBLIOGRAPHY

A F Wahlberg, Anders and Sjoberg, Lennart. "Risk Perception and the Media." *Journal of Risk Research* 3/1 (2000): 31-50.

A. Barreto, Raul. "Fossil Fuels, Alternative Energy and Economic Growth." *Economic Modelling* 75 (2018): 196-220.

A. Felix, Agbetuyi et al. "Wind Energy Potential in Nigeria." *International Electrical Engineering Journal* 3/1 (2012): 595-601.

A. Likhtarev, Ilya et al. "Chernobyl Accident: Retrospective and Prospective Estimates of External Dose of the Population of Ukraine." *Health Physics* 82/3 (2002): 290-303.

A. Rosa, Eugene et al. "Nuclear Waste: Knowledge Waste?." *Science* 329/5993 (2010): 762-763.

A. Schreurs, Miranda. "The Ethics of Nuclear Energy: Germany's Energy Politics after Fukushima." *The Journal of Social Science* 77 (2014): 9-29.

Abas, Naeem, Kalair, A. Raza and Khan, Nasrullah. "Review of Fossil Fuels and Future Energy Technologies." *Futures* 69 (2015): 31-49.

Adamantiades, Achilles and Kessides, Ioannis. "Nuclear Power for Sustainable Development: Current Status and Future Prospects." *Energy Policy* 37/12 (2009): 5149-5166.

Ağbulut, Ümit and Sarıdemir, Suat. "A General View to Converting Fossil Fuels to Cleaner Energy Source by Adding Nanoparticles." *International Journal of Ambient Energy* 42/13 (2021): 1569-1574.

Akyüz, Emrah. "Türkiye'de Çevresel Terör: PKK Terör Örgütü Örneği." *Ulusal Çevre Bilimleri Araştırma Dergisi* 34 (2020): 197-207.

Akyüz, Emrah. *Çevre Biliminin ABC'si: Yeni Başlayanlar İçin Çevre Sorunları ve Politikaları*. Ankara: Seçkin Yayıncılık, 2020.

Alagöz, Mehmet, Dilbaz Alacahan, Nur and Akarsu,Yağmur. "Petrol Fiyatlarının Makro Ekonomi Üzerindeki Etkisi - Ülke Karşılaştırmaları ile Panel Veri Analizi." *Karamanoğlu Mehmetbey Üniversitesi Sosyal ve Ekonomik Araştırmalar Dergisi* 19/33 (2017): 144-150.

Alashan, Sadık, Sen, Zekai and Fuat Toprak, Z. "Hydroelectric Energy Potential of Turkey: A Refined Calculation Method." *Arabian Journal for Science and Engineering* 41/4 (2016): 1511-1520.

Alexeeva, Victoria et al. "Measuring Employment Generated by the Nuclear Power Sector." No. NEA--7204. Organisation for Economic Co-Operation and Development, 2018.

Allen, Lucy et al. "Fossil Fuels and Water Quality," İçinde *the World's Water*, ed. Peter H. Gleick. Washington, DC: Island Press, 2012.

Allen, Todd et al. "Materials Challenges for Nuclear Systems." *Materials Today* 13/12 (2010): 14-23.

Alrikabi, Nada. "Renewable Energy Types." *Journal of Clean Energy Technologies* 2/1 (2014): 61-64.

Andrea Spadaro, Paola. "Climate Change, Environmental Terrorism, Eco-terrorism and Emerging Threats." *Journal of Strategic Security* 13/4 (2020): 58-80.

Apergis, Nicholas and E. Payne, James. "A Panel Study of Nuclear Energy Consumption and Economic Growth." *Energy Economics* 32/3 (2010): 545-549.

Apergis, Nicholas et al. "On the Causal Dynamics Between Emissions, Nuclear Energy, Renewable Energy, and Economic Growth." *Ecological Economics* 69/11 (2010): 2255-2260.

Atar, Emrah. "Managing Air Pollution in International Environmental Policies: Covid-19 as a Case to Demonstrate the Impact of Human on Environment." *Journal of Academic Value Studies* 7/2 (2021): 63-72.

Atar, Emrah. "Birleşmiş Milletler Dünya Gıda Programı'nın Evrensel Gıda Sistemi ve Dağıtımındaki Rolünün Analizi." *Disiplinlerarası Gıda Çalışmaları Dergisi* 1/1 (2021): 45-61.

Aven, Terje and Guikema, Seth. "On the Concept and Definition of Terrorism Risk." *Risk Analysis* 35/12 (2015): 2162-2171.

Aydın, İbrahim. "Balıkesir'de Rüzgâr Enerjisi." *Doğu Coğrafya Dergisi*, 18/29 (2013): 29-50.

Azar, Christian et al. "Carbon Capture and Storage from Fossil Fuels and Biomass–costs and Potential Role in Stabilizing the Atmosphere." *Climatic Change* 74/1 (2006): 47-79.

Azzuni, Abdelrahman and Breyer, Christian. "Energy Security and Energy Storage Technologies." *Energy Procedia* 155 (2018): 237-258.

B. Fridleifsson, Ingvar. "Geothermal Energy for the Benefit of the People." *Renewable and Sustainable Energy Reviews* 5/3 (2001): 299-312.

B.Moysich, Kirsten, J. Menezes, Ravi and M. Michalek, Arthur. "Chernobyl-related Ionising Radiation Exposure and Cancer Risk: An Epidemiological Review." *The Lancet Oncology* 3/5 (2002): 269-279.

Bahar Şanlı, Fatma and Tuna, Kadir. "Türkiye'de Petrol Tüketimi ile Ekonomik Büyüme Arasındaki İlişkinin Analizi." *Maliye ve Finans Yazıları* 102 (2014): 43-58.

Barbier, Enrico. "Geothermal Energy Technology and Current Status: An Overview." *Renewable and Sustainable Energy Reviews* 6/1-2 (2002): 3-65.

Barbir, Frano T., Veziroğlu, Nejat and Plass, H. J. "Environmental Damage due to Fossil Fuels Use." *International Journal of Hydrogen Energy* 15/10 (1990): 739-749

Barnaby, Frank. "The Effects of the Atomic Bombings of Hiroshima and Nagasaki." *Medicine, Conflict and Survival* 11/3 (1995): 1-9.

Bauera, Nico, J. Brechab, Robert and Luderer, Gunnar. "Economics of Nuclear Power and Climate Change Mitigation Policies." *Proceedings of the National Academy of Sciences* 109/42 (2012): 16805-16810.

Beck, Ulrich. *Ulrich Beck: Pioneer in Cosmopolitan Sociology and Risk Society*. Cham: Springer International Publishing, 2014.

Berkün, Mehmet, Aras, Egemen and Koç, Tuğçe. "Barajların ve Hidroelektrik Santrallerin Nehir Ekolojisi Üzerinde Oluşturduğu Etkiler." *Türkiye Mühendislik Haberleri* 452/6 (2008): 41-48.

Bian, Qi et al. "Risk Perceptions of Nuclear Energy, Climate Change, and Earthquake: How are They Correlated and Differentiated by Ideologies?." *Climate Risk Management* 32 (2021): 100297.

Bickerstaff, Karen. "Because We've Got History Here": Nuclear Waste, Cooperative Siting, and the Relational Geography of a Complex Issue."" *Environment and Planning A* 44/11 (2012): 2611-2628.

Bielecki, J. "Energy Security: Is the Wolf at the Door?." *The Quarterly Review of Economics and Finance* 42/2 (2002): 235-250.

Bilgen, Selcuk, Kaygusuz, Kamil and Sari, Ahmet. "Renewable Energy for a Clean and Sustainable Future." *Energy Sources* 26/12 (2004): 1119-1129.

Bodansky, David. *Nuclear Energy: Principles, Practices, and Prospects*. New York: Springer Science & Business Media, 2007.

Bruce, Gregor. "Definition of Terrorism Social and Political Effects." *Journal of Military and Veterans Health* 21/2 (2013): 26-30.

C. Ewing, Rodney, A. Whittleston, Robert and W. D. Yardley, Bruce. "Geological Disposal of Nuclear Waste: A Primer." *Elements* 12/4 (2016): 233-237.

C. Howard, David et al. "The Impact of Sustainable Energy Production on Land Use in Britain through to 2050." *Land Use Policy* 26 (2009): S284-S292.

C. Patterson, Walter. *Nuclear Power*. Harmondsworth: Penguin, 1976.

C. Turkenburg, Wim. "Renewable Energy Technologies," içinde *World Energy Assessment, Energy and the Challenge of Sustainability.* New York: UNDP, 2000.

C. Whitfield, Stephen et al. "The Future of Nuclear Power: Value Orientations and Risk Perception." *Risk Analysis: An International Journal* 29/3 (2009): 425-437.

Can Aşıkoğlu, Yakut and Tosunoğlu, Murat. "Rüzgâr Enerji Santral Sahalarındaki Kuşların Üreme Faaliyetlerinin Araştırılması." *Turkish Journal of Bioscience and Collections* 2/1 (2018): 12-26.

Can Şenel, Mahmut and Koç, Erdem. "Dünyada ve Türkiye'de Rüzgar Enerjisi Durumu-Genel Değerlendirme." *Mühendis ve Makina* 56/663 (2015): 46-56.

Castells, Manuel. "An Introduction to the Information Age." *City* 2/7 (1997): 6-16.

Ceribasi, Gokmen and Caliskan, Muhammed. "Short-And Long-term Prediction of Energy to be Produced in Hydroelectric Energy Plants of Sakarya Basin in Turkey." *Energy Sources, Part A: Recovery, Utilization, and Environmental Effects*, (2019): 1-16.

Çetin, Murat and Kürşat Demirci, Oğuz. "Erzincan'da Doğal Gaz Kullanımının Hava Kalitesine Etkisi." *Erzincan Üniversitesi Fen Bilimleri Enstitüsü Dergisi* 9/1 (2016): 8-18.

Chen, Beiwen et al. "Basic Psychological Need Satisfaction, Need Frustration, and Need Strength Across Four Cultures." *Motivation and Emotion*, 39/2 (2015): 216-236.

Chiari, Luca and Zecca, Antonio. "Constraints of Fossil Fuels Depletion on Global Warming Projections." *Energy Policy* 39/9 (2011): 5026-5034.

Covert, Thomas, Greenstone, Michael and R. Knittel, Christopher. "*Will We Ever Stop Using Fossil Fuels?*." *Journal of Economic Perspectives* 30/1 (2016): 117-38.

D. Arbuthnott, Katherine and Dolter Kahneman, Brett. "Escalation of Commitment to Fossil Fuels." *Ecological Economics* 89 (2013): 7-13.

D. Leslie, William. *Nuclear Medicine.* Boca Raton: CRC Press, 2003.

Dale, Michael, Krumdieck, Susan and Bodger, Pat. "Net Energy Yield from Production of Conventional Oil." *Energy Policy* 39/11 (2011): 7095-7102.

Dalla, Alessandra and Claudia Furlan, Valle. "Diffusion of Nuclear Energy in Some Developing Countries." *Technological Forecasting and Social Change* 81 (2014): 143-153.

Dannreuther, Roland. *Energy Security*. Cambridge: John Wiley & Sons, 2017.

Degueldre, Claude. “Uranium as a Renewable for Nuclear Energy.” *Progress in Nuclear Energy* 94 (2017): 174-186.

Demirbas, Ayhan. “Global Renewable Energy Projections.” *Energy Sources, Part B: Economics, Planning, and Policy* 4/2 (2009): 212-224.

Demirbaş, Ayhan. “Global Renewable Energy Resources.” *Energy Sources* 28/8 (2006): 779-792, 779.

Di Pascolia, Stefano, Femiab, Aldo and Luzzat, Tommaso. “Natural Gas, Cars and the Environment. A (relatively) ‘Clean’ and Cheap Fuel Looking for Users.” *Ecological Economics* 38/2 (2001): 179-189, 179.

Dincer, Ibrahim. “Renewable Energy and Sustainable Development: A Crucial Review.” *Renewable and Sustainable Energy Reviews* 4/2 (2000): 157-175.

Dittmar, Michael. “Nuclear Energy: Status and Future Limitations.” *Energy* 37/1 (2012): 35-40.

Dittmar, Michael. “The End of Cheap Uranium.” *Science of the Total Environment* 461: 792-798.

Doğan, Mesut. “Enerji Kullanımının Coğrafi Çevre Üzerindeki Etkileri,” *Marmara Coğrafya Dergisi* 23 (2011): 36-52.

Du, Xinming et al. “Transboundary Air Pollution from Coal-fired Power Generation.” *Journal of Environmental Management* 270 (2020): 1-9.

Duan, Xiaoli et al. “Household Fuel Use for Cooking and Heating in Chin`a: Results from the First Chinese Environmental Exposure-Related Human Activity Patterns Survey (CEERHAPS).” *Applied Energy* 136 (2014): 692-703.

E. Brockway, Paul et al. “Estimation of Global Final Stage Energy Return on Investment for Fossil Fuels with Comparison to Renewable Energy Sources.” *Nature Energy* 4/7 (2019): 612-621.

E. Kasperson, Roger et al. “Public Opposition to Nuclear Energy: Retrospect and Prospect.” *Science, Technology, & Human Values* 5/2 (1980): 11-23.

E. Maczulak, Anne. *Renewable Energy: Sources and Methods.* New York: Infobase Publishing, 2010.

Ebrahimi Sarcheshmeh, Elaheh, Bijani, Masoud and Sadighi, Hassan. “Adoption Behavior Towards the Use of Nuclear Technology in Agriculture: A causal Analysis.” *Technology in Society* 55 (2018): 175-182.

Efe Biresselioglu, Mehmet and Ozge Yumurtaci, Isik. “Evaluating the Nature of Terrorist Attacks on the Energy İnfrastructure: The Periodical Study for 1970-2011.” *International Journal of Oil, Gas and Coal Technology* 10/3 (2015): 325-341.

Ekener-Petersen, Elisabeth, Höglund, Jonas and Finnveden, Göran. "Screening Potential Social Impacts of Fossil Fuels and Biofuels for Vehicles." *Energy Policy* 73 (2014): 416-426.

Elizabeth Bell, Shannon and York, Richard. "Coal, Injustice, and Environmental Destruction: Introduction to the Special İssue on Coal and the Environment." *Organization & Environment* 25/4 (2012): 359-367.

Ellabban, Omar, Abu-Rub, Haitham and Blaabjerg, Frede. "Renewable Energy Resources: Current Status, Future Prospects and Their Enabling Technology." *Renewable and Sustainable Energy Reviews* 39 (2014): 748-764.

Emeklier, Bilgehan and Ergül, Nihal. "Petrolün Uluslararası İlişkilerdeki Yeri: Jeopolitik Teoriler ve Petropolitik." *Bilge Strateji* 2/3 (2010): 59-85.

Erlenbusch, Verena. "How (not) to Study Terrorism." *Critical Review of International Social and Political Philosophy* 17/4 (2014): 470-491.

Erol, Emine. "Türkiye'de Elektrik Enerjisinin Tarihi Gelişimi: 1902-2000." Yayınlanmamış Doktora Tezi, İstanbul Üniversitesi Sosyal Bilimler Enstitüsü, 2007.

F. de O. Falca˜o, Anto´nio. "Wave Energy Utilization: A Review of the Technologies." *Renewable and Sustainable Energy Reviews* 14/3 (2010): 899-918.

Fadaeenejad, M. et al. "New Approaches in Harnessing Wave Energy: With Special Attention to Small Islands." *Renewable and Sustainable Energy Reviews*, 29 (2014): 345-354.

Farzaneh-Gord, Mahmood, Hashemi, Shahram and Sadi, Meisam. "Energy Destruction in Iran's Natural Gas Pipe Line Network." *Energy Exploration & Exploitation* 25/6 (2007): 393-406.

Flynn, James et al. "Time to Rethink Nuclear Waste Storage." *Issues in Science and Technology* 8/4 (1992): 42-48.

Foster, Robert, Ghassemi, Majid and Cota, Alma. *Solar Energy: Renewable Energy and the Environment*. Boca Raton: CRC Press, 2009.

Fthenakis, Vasilis and Chul Kim, Hyung. "Land Use and Electricity Generation: A Life-cycle Analysis." *Renewable and Sustainable Energy Reviews* 13/6-7 (2009): 1465-1474.

Furlan, Claudia, Guidolin, Mariangela and Guseo, Renato. "Has the Fukushima Accident Influenced Short-term Consumption in the Evolution of Nuclear Energy? An Analysis of the World and Seven Leading Countries." *Technological Forecasting and Social Change* 107 (2016): 37-49.

G. Corlu, Canan et al. "Optimizing Energy Consumption in Transportation: Literature Review, Insights, and Research Opportunities." *Energies*, 13/5 (2020): 1115-1148.

G. Granqvist, Claes. "Solar Energy Materials." *Advanced Materials* 15/21 (2003): 1789-1803.

G. Olivier, Jos et al. "Recent Trends in Global Greenhouse Gas Emissions: Regional Trends 1970–2000 and Spatial Distributionof Key Sources in 2000." *Environmental Sciences* 2(/2-3 (2005): 81-99.

Gabriel, Sophie et al. "A Critical Assessment of Global Uranium Resources, Including Uranium in Phosphate Rocks, and the Possible Impact of Uranium Shortages on Nuclear Power Fleets," *Annals of Nuclear Energy* 58 (2013): 213-220.

Gao, Yi-Xuan et al. "Road Transport Energy Consumption in the G7 and BRICS: 1973-2010." *International Journal of Global Energy Issues* 38/4-6 (2015): 342-356.

Gielen, Dolf et al. "The Role of Renewable Energy in the Global Energy Transformation." *Energy Strategy Reviews* 24 (2019): 38-50, 38.

Giurco, Damien et al. "Requirements for Minerals and Metals for 100% Renewable Scenarios." İçinde *Achieving the Paris Climate Agreement Goals*, ed. Sven Teske. Cham: Springer, Cham, 2019.

Gökçe, Cem. "Önemli Bir Enerji Girdisi Olan Petrolün Ekonomik Kalkınma Sürecindeki Rolü." *Afyon Kocatepe Üniversitesi İktisadi ve İdari Bilimler Fakültesi Dergisi* 16/1(2015): 143-154.

Goswami, Sribas. "Impact of Coal Mining on Environment." *European Researcher* 92/3 (2015):185-196.

Green, Richard and Staffell, Iain. "Electricity in Europe: Exiting Fossil Fuels?." *Oxford Review of Economic Policy* 32/2 (2016): 282-303.

Groot, De et al. "Values, Perceived Risks and Benefits, and Acceptability of Nuclear Energy." *Risk Analysis: An International Journal* 33/2 (2013): 307-317.

Gross, Robert, Leach, Matthew and Bauen, Ausilio. "Progress in Renewable Energy." *Environment International* 29/1 (2003): 105-122.

Guerra, Olivia, Laure, Santin and Henk Vissche, Itard. "The Effect of Occupancy and Building Characteristics on Energy Use for Space and Water Heating in Dutch Residential Stock." *Energy and Buildings*, 41/11 (2009), 1223-1232.

Guo, Wenkai et al. "Ambient $PM_{2.5}$ and Related Health Impacts of Spontaneous Combustion of Coal and Coal Gangue." *Environmental Science & Technology*, 55/9 (2021), 5763-5771.

Gustavsson, Leif et al. "Climate Change Effects of Forestry and Substitution of Carbon-intensive Materials and Fossil Fuels." *Renewable and Sustainable Energy Reviews* 67 (2017): 612-624.

H. Dale, Virginia, A. Efroymson, Rebecca and L. Kline, Keith. "The Land Use Climate Change Energy Nexus." *Landscape Ecology* 26/6 (2011): 755-773.

H. Gleick, Peter. "Environmental Consequences of Hydroelectric Development: The Role of Facility Size and Type." *Energy* 17/8 (1992): 735-747.

H. Kuklinski, James, S. Metlay, Daniel and D. Kay., W. "Citizen Knowledge and Choices on the Complex Issue of Nuclear Energy." *American Journal of Political Science* 26/4 (1982): 615-642.

Hachiya, Michihiko. *Hiroshima Diary: The Journal of a Japanese Physician, August 6-September 30, 1945*. Chapel Hill: UNC Press Books, 2011.

Häfele, W. "Hypotheticality and the New Challenges: The Pathfinder Role of Nuclear Energy," *Minerva* 12/3 (1974): 303-322.

Haily Dalvi, Vishwanath, V. Panse, Sudhir and B. Joshi, Jyeshtharaj. "Solar Thermal Technologies as a Bridge From Fossil Fuels to Renewables." *Nature Climate Change* 5/11 (2015): 1007-1013.

Harjanne, Atte and M. Korhonen, Janne. "Abandoning the Concept of Renewable Energy." *Energy Policy* 127 (2019): 330-340.

Hassan, Ather et al. "Monetization of the Environmental Damage Caused by Fossil Fuels." *Environmental Science and Pollution Research* 28/17 (2021): 21204-21211.

Hayhoe, Katharine et al. "Substitution of Natural Gas for Coal: Climatic Effects of Utility Sector Emissions." *Climatic Change* 54/1 (2002): 107-139.

Helms, Hinrich and Lambrecht, Udo. "The Potential Contribution of Light-Weighting to Reduce Transport Energy Consumption." *International Journal of Life Cycle Assessment* 12/1 (2007): 58-64.

Henrich, Edmund et al. "The Role of Biomass in a Future World Without Fossil Fuels," *Chemie Ingenieur Technik* 87/12 (2015): 1667-1685.

Hiyama, Atsuki et al. "The Biological Impacts of the Fukushima Nuclear Accident on the Pale Grass Blue Butterfly." *Scientific Reports* 2/1 (2012): 1-11.

Hocke, Peter and Renn, Ortwin. "Concerned Public and the Paralysis of Decision-making: Nuclear Waste Management Policy in Germany." *Journal of Risk Research* 12/7-8 (2009): 921-940.

Hommerich, Carola. "Trust and Subjective Well-being After the Great East Japan Earthquake, Tsunami and Nuclear Meltdown: Preliminary Results." *International Journal of Japanese Sociology* 21/1 (2012): 46-64.

Höök, Mikael and Tang, Xu. "Depletion of Fossil Fuels and Anthropogenic Climate Change—A Review." *Energy Policy* 52 (2013): 797-809.

Hore-Lacy, Ian. *Nuclear Energy in the 21st Century*. London: World Nuclear University Press, 2010.

Horrigan, Leo, S. Lawrence, Robert and Walker P., Polly. "How Sustainable Agriculture Can Address the Environmental and Human Health Harms of Industrial Agriculture." *Environmental Health Perspectives*, 110/5 (2002): 445-456.

Hosgood, H. Dean et al. "Household Coal Use and Lung Cancer: Systematic Review and Meta-analysis of Case–Control Studies, with an Emphasis on Geographic Variation." *International journal of Epidemiology*, 40/3 (2011): 719-728.

HoSong, Jin. "An Assessment on the Environmental Contamination Caused by the Fukushima Accident." *Journal of Environmental Management* 206 (2018): 846-852.

Huang, Lei et al. "The Changing Risk Perception Towards Nuclear Power in China after the Fukushima Nuclear Accident in Japan." *Energy Policy* 120 (2018): 294-301.

Hubbert, M. King. "Energy from Fossil Fuels." *Science* 109/2823 (1949): 103-109.

I. Joensuu, Oiva. "Fossil Fuels as a Source of Mercury Pollution." *Science* 172/3987 (1971): 1027-1028.

İlkiliç, Cumali. "Wind Energy and Assessment of Wind Energy Potential in Turkey." *Renewable and Sustainable Energy Reviews*, 16/2 (2012): 1165-1173.

Imran Khan, Muhammad. "Policy Options for the Sustainable Development of Natural Gas as Transportation Fuel." *Energy Policy* 110 (2017): 126-136.

Inman, Mason. The True Cost of Fossil Fuels. *Scientific American* 308/4 (2013): 58-61.

Isabel Blanco, Maria and Rodrigues, Glória. "Direct Employment in the Wind Energy Sector: An EU Study." *Energy Policy* 37/8 (2009): 2847-2857.

J. Chew, Kenneth. "The Future of Oil: Unconventional Fossil Fuels." *Philosophical Transactions of the Royal Society A: Mathematical, Physical and Engineering Sciences* 372/2006 (2014): 1-34.

J. Cleveland, Cutler. "The Direct and Indirect Use of Fossil Fuels and Electricity in USA Agriculture, 1910–1990." *Agriculture, Ecosystems & Environment* 55/2 (1995): 111-121.

J. Hager, Tiffany and Morawicki, Ruben. "Energy Consumption During Cooking in the Residential Sector of Developed Nations: A Review." *Food Policy* 40 (2013): 54-63.

J. Madia, W. "The 21st Century Nuclear Park: A Source of Energy, Water, Food, and Jobs." *International Journal of Nuclear Energy Science and Technology* 1/1 (2004): 7-19.

J. Otway, Harry, Maurer, Dagmar and Thomas, Kerry. "Nuclear Power: The Question of Public Acceptance," *Futures* 10/2 (1978): 109-118.

J. Scheidt, Rick. "The Babushkas of Chernobyl." *The Gerontologist* 58/1 (2018): 193-195.

J. Zinkle, Steven and S. Was, Gary. "Materials Challenges in Nuclear Energy." *Acta Materialia* 61/3 (2013): 735-758.

János Besenyő, Col. "Inferno Terror: Forest Fires as the New form of Terrorism." *Terrorism and Political Violence* 31/6 (2019): 1229-1241.

Johansson, Olof and Schipper, Leo. "Measuring the Long-Run Fuel Demand of Cars: Separate Estimations of Vehicle Stock, Mean Fuel İntensity, and Mean Annual Driving Distance." *Journal of Transport Economics and Policy*, 31/3 (1997): 277-292.

Johnston, I. W. et al. "Emerging Geothermal Energy Technologies." *KSCE Journal of Civil Engineering* 15/4 (2011): 643-653.

José Vidal-Amaro, Juan et al. "Optimal Energy Mix for Transitioning from Fossil Fuels to Renewable Energy Sources–The Case of the Mexican Electricity System." *Applied Energy* 150 (2015): 80-96.

K. Sovacool, Benjamin. "Valuing the Greenhouse Gas Emissions from Nuclear Power: A Critical Survey." *Energy Policy*, 36/8 (2008): 2950-2963.

K. Yeo, Sara et al. "Partisan Amplification of Risk: American Perceptions of Nuclear Energy Risk in the Wake of the Fukushima Daiichi Disaster." *Energy Policy* 67 (2014): 727-736.

K. Biswas, Asit. "Impacts of Hydroelectric Development on the Environment." *Energy Policy* 10/4 (1982): 349-354.

K.Hopke, Philip. "Contemporary Threats and Air Pollution." *Atmospheric Environment* 43/1 (2009): 87-93.

Kaan Yeşilyurt, Muhammet et al. "A Scrutiny Study on Wave Energy Potential and Policy in Turkey." *Periodicals of Engineering and Natural Sciences* 5/3 (2017): 286-297.

Kåberger, Tomas. "Progress of Renewable Electricity Replacing Fossil Fuels." *Global Energy Interconnection* 1/1 (2018): 48-52.

Kaltschmitt, Martin, Streicher, Wolfgang and Wiese, Andreas (eds.). *Renewable Energy: Technology, Economics and Environment*. New York: Springer Science & Business Media, 2007.

Karadeniz, Vedat, Akpınar, Erdal and Başıbüyük, Adem. "Nehir Tipi Hidroelektrik Santraller ve Çevresel Etkileri (Reşadiye Hidroelektrik Santralleri Örneği)," *Doğu Coğrafya Dergisi* 16/26 (2011): 95-114.

Kerimray, Aiymgul et al. "Coal Use for Residential Heating: Patterns, Health Implications and Lessons Learned." *Energy for Sustainable Development*, 40 (2017): 19-30.

Kerr Casper, Julie. *Fossil Fuels and Pollution: The Future of Air Quality*. New York: Infobase Publishing, 2010.

Kim, Hyungjin, Park, Chul and Joon Kwon, Oh. "Conceptual Design of the Space Disposal System for the Highly Radioactive Component of the Nuclear Waste." *Energy* 115 (2016): 155-168.

Kit Lam, Wah. "Lung Cancer in Asian Women—The Environment and Genes." *Respirology*, 10/4 (2005): 408-417.

Kruyt, Bert et al. "Indicators for Energy Security." *Energy Policy* 37/6 (2009): 2166-2181.

Kuang, Yonghong et al. "A Review of Renewable Energy Utilization in Islands." *Renewable and Sustainable Energy Reviews* 59 (2016): 504-513.

L. Arnold, Jeffrey et al. "A Proposed Universal Medical and Public Health Definition of Terrorism." *Prehospital and Disaster Medicine* 18/2 (2003): 47-52.

L. Chalecki, Elizabeth. "A New Vigilance: Identifying and Reducing the Risks of Environmental Terrorism," *Global Environmental Politics* 2/1 (2002): 46-64.

L. Klass, Donald. Biomass for Renewable Energy, Fuels, and Chemicals. London: Elsevier, 1998.

L. Koerner, Cassandra. "Media, Fear, and Nuclear Energy: A Case Study." *The Social Science Journal* 51/2 (2014): 240-249.

L. Simon, Steven et al. "Estimated Radiation Doses Received by New Mexico Residents from the 1945 Trinity Nuclear Test." *Health Physics* 119/4 (2020): 428-477.

L. Toth, Ferenç and Rogner, Hans-Holger. "Oil and Nuclear Power: Past, Present, and Future." *Energy Economics* 28/1 (2006): 1-25.

Lakatos, L., Hevessy, G. and Kovács, J. "Advantages and Disadvantages of Solar Energy and Wind-power Utilization." *World Futures* 67/6 (2011): 395-408.

Laxmi Pandey, Vijay and Chaubal, Aditi. "Comprehending Household Cooking Energy Choice in Rural India." *Biomass and Bioenergy*, 3511 (2011): 4724-4731.

Lazkano, Itziar, Nøstbakken, Linda and Pelli, Martino. "From Fossil Fuels to Renewables: The Role of Electricity Storage." *European Economic Review* 99 (2017): 113-129.

Le Page, Michael. "World Without Fossil Fuels." *New Scientist*, 224/2991 (2014): 34-39.

Lehmann, Bernd. "Uranium Ore Deposits." *Reviews in Economic Geology* 2 (2008): 16-26.

Lehr, Ulrike et al. "Renewable Energy and Employment in Germany." *Energy Policy* 36/1 (2008): 108-117.

Lenzen, Manfred. "Life Cycle Energy and Greenhouse Gas Emissions of nuclear Energy: A Review." *Energy Conversion and Management* 49/8 (2008): 2178-2199.

Lewtas, Joellen. "Air Pollution Combustion Emissions: Characterization of Causative Agents and Mechanisms Associated with Cancer, Reproductive, and Cardiovascular Effects." *Mutation Research/Reviews in Mutation Research* 636/1-3 (2007): 95-133.

Li, Ming-Jia et al. "The Development Technology and Applications of Supercritical CO_2 power Cycle in Nuclear Energy, Solar Energy and Other Energy İndustries." *Applied Thermal Engineering* 126 (2017): 255-275.

López, Iraide et al. "Review of Wave Energy Technologies and the Necessary Power-equipment." *Renewable and Sustainable Energy Reviews* 27 (2013): 413-434.

M. Balmaceda, Margarita. "Differentiation, Materiality, and Power: Towards a Political Economy of Fossil Fuels." *Energy Research & Social Science* 39 (2018): 130-140.

M. Castleden, William et al. "The Mining and Burning of Coal: Effects on Health and the Environment." *The Medical Journal of Australia* 195/6 (2011): 333-335.

M. Fthenakis, Vasilis and Chul Kim, Hyung. "Greenhouse Gas Emissions from Solar Electric and Nuclear Power: A life Cycle Study." *Energy Policy* 35/4 (2007): 2549-2557

M. Janković, Marija et al. "Study of Radioactivity in Environment around Power Plants Tent a and Kolubara due to Coal Burning for 2015." *RAD Conference Proceedings* 1 (2016): 84-89.

M. Kammen, Daniel. "The Rise of Renewable Energy." *Scientific American* 295/3 (2006): 84-93.

M. Macfarlane, Allison and Miller, Marvin. “Nuclear Energy and Uranium Resources.” *Elements* 3/3 (2007): 185-192.

M. Mochizuki, Mike. “Japan Tests the Nuclear Taboo.” *Nonproliferation Review* 14/2 (2007): 303-328.

M. Thompson, Tammy et al. “Modeling to Evaluate Contribution of Oil and Gas Emissions to Air Pollution.” *Journal of the Air & Waste Management Association* 67/4 (2017): 445-461.

Maharika, Michael and Fischhoff, Baruch. “Public Views of Using Nuclear Energy Sources in Space Missions.” *Space Policy* 9/2 (1993): 99-108.

Maharikl, Michael and Fischhof, Baruch. “The Risks of Using Nuclear Energy Sources in Space: Some Lay Activists’ Perceptions.” *Risk Analysis* 12/3 (1992): 383-392.

Manieniyan, V., Thambidurai, M. and Selvakumar, R. “Study on Energy Crisis and the Future of Fossil Fuels.” *Proceedings of SHEE* 10 (2009): 2234-3689.

Markard, Jochen et al. “Destined for Decline? Examining Nuclear Energy from a Technological Innovation Systems Perspective.” *Energy Research & Social Science* 67 (2020): 1-19.

Mendel, Toby. “Freedom of Information as an Internationally Protected Human Right.” *Comparative Media Law Journal* 1/1 (2003): 39-70.

Mirchi, Ali et al. “World Energy Balance Outlook and OPEC Production Capacity: Implications for Global Oil Security” *Energies* 5/8 (2012): 2626-2651.

ML Moreira, João, Gallinaro, Bruno and Carajilescov, Pedro. “Construction Time of PWRs.” *Energy Policy* 55 (2013): 531-542.

Mohammad Bagher, Askari et al. “Hydroelectric Energy Advantages and Disadvantages.” *American Journal of Energy Science*, 2/2 (2015): 17-20.

Montel, Jean-Marc. “Minerals and Design of New Waste forms for Conditioning Nuclear Waste.” *Comptes Rendus Geoscience* 343/2-3 (2011): 230-236.

Montzka, S. A., Dlugokencky E J. and Butler, J. H. “Non-CO2 Greenhouse Gases and Climate Change. *Nature* 476/7358 (2011): 43-50.

Morris, Eric. “From Horse Power to Horsepower.” *Access Magazine*, 1/30 (2007), 2-10.

Mukhopadhyay, Kakali and Forssell, Osmo. “An Empirical Investigation of Air Pollution from Fossil Fuel Combustion and Its Impact on Health in India During 1973–1974 to 1996–1997.” *Ecological Economics* 55/2 (2005): 235-250.

Muradov, Nazim. "Hydrogen via Methane Decomposition: An Application for Decarbonization of Fossil Fuels." *International Journal of Hydrogen Energy* 26/11 (2001): 1165-1175.

Murray, Raymond and E. Holbert, Keith. *Nuclear Energy: An İntroduction to the Concepts, Systems, and Applications of Nuclear Processes*. Waltham, Oxford: Elsevier, 2014.

Naci Bayraç, Hüseyin. "Küresel Petrol Piyasasında Fiyat Oluşumu ve Ekonomik Etkileri." *Yalova Sosyal Bilimler Dergisi* 9/19 (2019): 44-59.

Nakicenovic, Nebojsa. "The Automobile Road to Technological Change: Diffusion of the Automobile as a Process of Technological Substitution." *Technological Forecasting and Social Change*, 29/4 (1986): 309-340.

Nambiar, Deepika. "The Impact of Online Learning During COVID-19: Students' and Teachers' Perspective." *The International Journal of Indian Psychology* 8/2 (2020): 783-793.

New Daggett, Cara. *The Birth of Energy: Fossil Fuels, Thermodynamics and the Politics of Work.* Durham: Duke University Press, 2019.

Nicoletti, Giovanni et al. "A Technical and Environmental Comparison Between Hydrogen and Some Fossil Fuels." *Energy Conversion and Management* 89 (2015): 205-213.

Nield, Ted. *Supercontinent: Ten Billion Years in the Life of Our Planet.* Cambridge: Harvard University Press, 2007.

O. Kaplanab, Jed, M. Krumhardta, Kristen and Zimmermann, Niklaus. "The Prehistoric and Preindustrial Deforestation of Europe." *Quaternary Science Reviews* 28/27-28 (2009): 3016-3034.

O'Lear, Shannon. "Environmental Terrorism: A Critique." *Geopolitics* 8/3 (2003): 127-150.

Olofsson, Anna and Öhman, Susanna. "Vulnerability, Values and Heterogeneity: One Step Further to Understand Risk Perception and Behaviour." *Journal of Risk Research* 18/1 (2015): 2-20.

Oral, Faruk, Behçet, Rasim and Aykut, Kadir. "Hidroelektrik Santral Rezervuar Verilerinin Enerji Üretimi Amaçlı Değerlendirilmesi." *Bitlis Eren Üniversitesi Fen Bilimleri Dergisi* 6/2 (2017): 29-38.

Ortaş, İbrahim. "Bilgi ve İletişim Çağında Bilimsel Bilgiye Erişimin Önemi ve Türkiye'nin Bilgiye Erişim Potansiyeli." *Türk Kütüphaneciliği*, 32/3 (2018): 223-232.

Osman, Uzun. "Hidro Elektrik Santraller (HES) ve Çevresel Etki Değerlendirmesi (ÇED) Düzce Örneği." *Düzce Üniversitesi Orman Fakültesi Ormancılık Dergisi* 7/2 (2011): 1-13.

Otway, Harry and Thomas, Kerry. "Reflections on Risk Perception and Policy 1, 2." *Risk Analysis* 2/2 (1982): 69-82.

Ourisson, Guy, Pierre Albrecht, and Michel Rohmer. "The Microbial Origin of Fossil Fuels," *Scientific American* 251/2 (1984): 44-51.

Özen, Ahmet, Ünsal Şaşmaz, Mahmut and Bahtiyar, Ercan. "Türkiye'de Yeşil Ekonomi Açısından Yenilenebilir Bir Enerji Kaynağı: Rüzgâr Enerjisi." *Karamanoğlu Mehmetbey Üniversitesi Sosyal ve Ekonomik Araştırmalar Dergisi*, 17/28 2015: 85-93.

Özer Paşalıoğlu, Tülün. "Kentsel Alanlarda Görsel Kirlilik: Tekirdağ Örneği." Master's thesis, Trakya Üniversitesi Fen Bilimleri Enstitüsü, 2007.

P. Barke, Richard and C. Jenkins-Smith, Hank. "Politics and Scientific Expertise: Scientists, Risk Perception, and Nuclear Waste Policy." *Risk Analysis* 13/4 (1993): 425-439.

P. Fletcher, George. "The Indefinable Concept of Terrorism." *Journal of International Criminal Justice* 4/5 (2006): 894-911.

P. Kuvlesky, William et al. "Wind Energy Development and Wildlife Conservation: Challenges and Opportunities." *The Journal of Wildlife Management*, 71/8 (2007): 2487-2498.

P. Sims, Geoffrey. "Hydroelectric Energy." *Energy Policy*, 19/8 (1991): 776-786.

Pai, Sandeep et al. "Meeting Well-below 2 C Target Would Increase Energy Sector Jobs Globally." *One Earth* 4/7 (2021): 1026-1036.

Panwar, N. L., Kaushik, S. C. and Kothari, Surendra. "Role of Renewable Energy Sources in Environmental Protection: A Review." *Renewable and Sustainable Energy Reviews* 15/3 (2011): 1513-1524.

Patel, Nivedita. "Environmental and Economical Effects of Fossil Fuels." *Journal of Recent Research in Engineering and Technology* 1 (2014): 1-12.

Penalba, Markel, Giorgi, Giussepe and V. Ringwood, John. "Mathematical Modelling of Wave Energy Converters: A Review of Nonlinear Approaches." *Renewable and Sustainable Energy Reviews* 78 (2017): 1188-1207.

Pershagen, B. and Bowen, M. *Light Water Reactor Safety*. Oxford: Pergamon Press, 2013.

Petrosyants, A. "The Soviet Union and the Development of Nuclear Power." *IAEA Bulletin* 28/3 (1986): 5-8.

Poinssot, Ch et al. "Assessment of the Environmental Footprint of Nuclear Energy Systems. Comparison between Closed and Open Fuel Cycles." *Energy* 69 (2014): 199-211.

Prati, Gabriele and Zani, Bruna. "The Effect of the Fukushima Nuclear Accident on Risk Perception, Antinuclear Behavioral İntentions, Attitude, Trust, Environmental Beliefs, and Values." *Environment and Behavior* 45/6 (2013): 782-798.

Primoratz, Igor. "What is Terrorism?." *Journal of Applied Philosophy* 7/2 (1990): 129-138.

Protásio, de Paula et al. "Brazilian Lignocellulosic Wastes for Bioenergy Production: Characterization and Comparison with Fossil Fuels," *BioResources* 8/1 (2013): 1166-1185.

R. Bohi, Douglas and A. Toman, Michael. *The Economics of Energy Security*. Massachusetts: Kluwer Academic Publishers, 2012.

R. Bull, Stanley. "Renewable Energy Today and Tomorrow." *Proceedings of the IEEE* 89/8 (2001): 1216-1226.

R. Timilsina, Govinda, Csordás, Stefan and Mevel, Simon. "When Does a Carbon Tax on Fossil Fuels Stimulate Biofuels?." Ecological Economics 70/12 (2011): 2400-2415.

R. Upadhyaya, Belle and W. Kerlin, Thomas. *Dynamics and Control of Nuclear Reactors*. London: Academic Press, 2019.

R. Watson, T. et al. "Greenhouse Gases: Sources and Sinks." *Climate Change* 92 (1992): 25-46.

Rachkov, V. I. et al. "From the First Nuclear Power Plant to Fourth-generation Nuclear Power Installations [On the 60th Anniversary of the World's First Nuclear Power Plant]." *Thermal Engineering* 61/5 (2014): 327-336.

Ramanathan, Ramakrishnan. "A Holistic Approach to Compare Energy Efficiencies of Different Transport Modes." *Energy Policy*, 28/11 (2000): 743-747.

Rehm, Jürgen, W. Lachenmeier, Dirk and Room, Robin. "Why Does Society Accept a Higher Risk for Alcohol than for Other Voluntary or Involuntary Risks?." *Bmc Medicine* 12/1 (2014): 1-6.

Rempel, Hilmar. "Geographical Distribution of Oil and Natural Gas Deposits-Different Means of Transportation to the Consumption Centers." In Pipeline Technology 2006 Conference Papers, Hannover, Germany. 2006.

Renn, Ortwin. "The Role of Risk Perception for Risk Management." *Reliability Engineering & System Safety* 59/1 (1998): 49-62.

Reza Lotfalipour, Mohammad, Ali Falahi, Mohammad and Ashena, Malihe. "Economic Growth, CO2 Emissions, and Fossil Fuels Consumption in Iran." *Energy* 35/12 (2010): 5115-5120.

Rodriguez, C. et al. "Deep-Burn: Making Nuclear Waste Transmutation Practical." *Nuclear Engineering and Design* 222/2-3 (2003): 299-317.

Rodríguez-Garzón, Ignacio et al. "Voluntary and Involuntary Risk Acceptance: A Case Study of Firefighters." *Safety Science* 142 (2021): 105394.

Rothwell, Geoffrey and van der Zwaan, Bob. "Are Light-water Reactor Energy Systems Sustainable?." *The Journal of Energy and Development* 29/1 (2003): 65-79.

S. Arutyunov, Vladimir and V. Lisichkin, Georgiy. "Energy Resources of the 21st Century: Problems and Forecasts. Can Renewable Energy Sources Replace Fossil Fuels?." *Russian Chemical Reviews* 86/8 (2017): 777-804.

S. Dresselhaus, Mildred and L. Thomas, I. "Alternative Energy Technologies." *Nature* 414/6861 (2001): 332-337.

Ş. Ediger, Volkan and Kentel, Elçin. "Renewable Energy Potential as an Alternative to Fossil Fuels in Turkey." *Energy Conversion and Management* 40/7 (1999): 743-755.

Ş. Ediger, Volkan. "An Integrated Review and Analysis of Multi-energy Transition from Fossil Fuels to Renewables." *Energy Procedia* 156 (2019): 2-6.

S. Emelyanov, V. "Nuclear Energy in the Soviet Union." *Bulletin of the Atomic Scientists* 27/9 (1971): 38-41.

S. Hecker, Siegfried. "Preventing Nuclear Weapon Proliferation as Nuclear Power Expands." *MRS bulletin* 33/4 (2008): 340-342.

S. Masnadi, Mohammad et al.. "From Fossil Fuels Towards Renewables: Inhibitory and Catalytic Effects on Carbon Thermochemical Conversion During Co-Gasification of Biomass with Fossil Fuels." *Applied Energy* 140 (2015): 196-209.

Saaty, T. L. and Gholamnezhad, H. "High-level Nuclear Waste Management:Analysis of Options." *Environment and Planning B: Planning and Design* 9/2 (1982): 181-196, 181.

Saha, Sabhasachi et al. "Rooppur Nuclear Power Plant: Current Status & Feasibility." *Strojnícky časopis-Journal of Mechanical Engineering* 68/3 (2018): 167-182.

Saidur, R. et al. "Environmental Impact of Wind Energy." *Renewable and Sustainable Energy Reviews* 15/5 (2011): 2423-2430.

Salvatores, Massimo. "Nuclear Fuel Cycle Strategies Including Partitioning and Transmutation." *Nuclear Engineering and Design* 235/7 (2005): 805-816, 805.

Sathre, Roger and Leif Gustavssonö "Time-dependent Climate Benefits of Using Forest Residues to Substitute Fossil Fuels," *Biomass and Bioenergy* 35/7 (2011): 2506-2516.

Saunders, P. "The Direct and Indirect Effects of Unemployment on Poverty and Inequality." *Australian Journal of Labour Economics* 5/4 (2002): 507-529.

Shafiee, Shahriar and Topal, Erkan. "An Econometrics View of Worldwide Fossil Fuel Consumption and the Role of US." *Energy Policy* 36/2 (2008): 775-786.

Shahzad, Umair. "The Need for Renewable Energy Sources." *Energy* 2 (2012): 16-18.

Shamshad, Ahmad, M. H., Fulekar and Bhawana, Pathak. "Impact of Coal Based Thermal Power Plant on Environment and Its Mitigation Measure." *International Research Journal of Environment Sciences* 1/4 (2012): 60-64.

Silje Kristiansen et al., "Risk Perception of Nuclear Energy after Fukushima: Stability and Change in Public Opinion in Switzerland," *International Journal of Public Opinion Research* 30/1 (2018): 24-50.

Simpson, Adam. "The Environment–Energy Security Nexus: Critical Analysis of an Energy 'Love Triangle'in Southeast Asia." *Third World Quarterly* 28/3 (2007): 539-554.

Sjöberg, Lennart. "Risk Perception: Experts and the Public." *European Psychologist* 3/1 (1998): 1-12.

Slovic, Paul, Fischhoff, Baruch and Lichtenstein, Sarah. "Why Study Risk Perception?." *Risk Analysis* 2/2 (1982): 83-93, 85.

Smallwood, K. Shawn. "Comparing Bird and Bat Fatality-Rate Estimates Among North American Wind-energy Projects." *Wildlife Society Bulletin* 37/1 (2013): 19-33.

Speirs, Jamie, McGlade, Christophe and Slade, Raphael. "Uncertainty in the Availability of Natural Resources: Fossil Fuels, Critical Metals and Biomass." *Energy Policy* 87 (2015): 654-664.

Spence, Martin. "Soviet Power: Nuclear Energy in the USSR." *Capital & Class* 7/3 (1983): 87-96.

Strandberg, Urban and Andrén, Mats. "Nuclear Waste Management in a Globalised World." *Journal of Risk Research* 12/7-8 (2009): 879-895.

Su, Po-An and Karney, d Bryan. "Micro Hydroelectric Energy Recovery in Municipal Water Systems: A Case Study for Vancouver." *Urban Water Journal* 12/8 (2015): 678-690.

Sun, Xiaoqi, Luo, Huimin and Dai, Sheng. "Ionic Liquids-based Extraction: A Promising Strategy for the Advanced Nuclear Fuel Cycle." *Chemical Reviews* 112/4 (2012): 2100-2128.

T. Huber, Matthew and McCarthy, James. "Beyond the Subterranean Energy Regime? Fuel, Land Use and the Production of Space." *Transactions of the Institute of British Geographers* 42/4 (2017): 655-668.

T. Huber, Matthew. "Energizing Historical Materialism: Fossil Fuels, Space and the Capitalist Mode of Production." *Geoforum* 40/1 (2009): 105-115.

T. Todorov, Peter and Nikolova Ilieva, Elica. Contamination with Uranium from Natural and Antropological Sources. *Romanian Journal of Physics* 51/1-2 (2006): 27-34.

Temurçin, Kadir and Aliağaoğlu, Alpaslan. "Nükleer Enerji ve Tartışmalar Işığında Türkiye'de Nükleer Enerji Gerçeği." *Coğrafi Bilimler Dergisi* 1/2 (2003): 25-39.

Tomonaga, Masao. "The Atomic Bombings of Hiroshima and Nagasaki: A Summary of the Human Consequences, 1945-2018, and Lessons for Homo Sapiens to end the Nuclear Weapon Age." *Journal for Peace and Nuclear Disarmament* 2/2 (2019): 491-517.

Tyrrell, Toby, G. Shepherd, John and Castle, Stephanie. "The Long-term Legacy of Fossil Fuels." *Tellus B: Chemical and Physical Meteorology* 59/4 (2007): 661-672.

Urbinato, David. "London's Historic" Pea-soupers." (Smog in London, England)." *EPA Journal* 20/1-2 (1994): 44-45.

Vaillancourt, Kathleen et al. "The Role of Nuclear Energy in Long-term Climate Scenarios: An Analysis with the World-TIMES Model." *Energy Policy* 36/7 (2008): 2296-2307.

Valero, Alicia and Valero, Antonio. "What are the Clean Reserves of Fossil Fuels?." *Resources, Conservation and Recycling* 68 (2012): 126-131.

Verbruggen, Aviel. "Renewable and Nuclear Power: A Common Future?." *Energy Policy* 36/11 (2008): 4036-4047.

von Wehrden, Henrik et al. "Consequences of Nuclear Accidents for Biodiversity and Ecosystem Services." *Conservation Letters* 5/2 (2012): 81-89.

W. Beck, Peter. "Nuclear Energy in the Twenty-first Century: Examination of a Contentious Subject." *Annual Review of Energy and the Environment* 24/1 (1999): 113-137.

W. Brook, Barry and J. A. Bradshaw, Corey. "Key Role for Nuclear Energy in Global Biodiversity Conservation." *Conservation Biology* 29/3 (2015): 702-712.

W. Crabtree, George and S. Lewis, Nathan. "Solar Energy Conversion." *Physics Today*, 60/3 (2007): 37-42.

W. Davis, Lucas. "Prospects for Nuclear Power." *Journal of Economic Perspectives* 26/1 (2012): 49-66.

W. Golay, Michael and E. Todreas, Neil. "Advanced Light-water Reactors." *Scientific American* 262/4 (1990): 82-89.

W. Russell Alexander and Linda McKinley, eds. *Deep Geological Disposal of Radioactive Waste.* Oxford: Elsevier, 2011.

W. Slimak, Michael and Dietz, Thomas. "Personal Values, Beliefs, and Ecological Risk Perception." *Risk Analysis* 26/6 (2006): 1689-1705.

Wah Ang, Beng, Lim Choong, Wei and Sheng Ng, Tsan. "Energy Security: Definitions, Dimensions and Indexes." *Renewable and Sustainable Energy Reviews* 42 (2015): 1077-1093.

Walker, Gordon. "Energy, Land Use and Renewables: A Changing Agenda." *Land Use Policy* 12/1 (1995): 3-6.

Wang, Shanyong et al. "Public Perceptions and Acceptance of Nuclear Energy in China: The Role of Public Knowledge, Perceived Benefit, Perceived Risk and Public Engagement." *Energy Policy* 126 (2019): 352-360.

Wang, Shifeng and Wang, Sicong. "Impacts of Wind Energy on Environment: A Review." *Renewable and Sustainable Energy Reviews* 49 (2015): 437-443.

Wang, Yuanfeng et al. "Transport Energy Consumption and Saving in China." *Renewable and Sustainable Energy Reviews* 29 (2014): 641-655, 641.

Wei, Max, Patadia, Shana and M. Kammen, Daniel. "Putting Renewables and Energy Efficiency to Work: How Many Jobs Can the Clean Energy Industry Generate in the US?." *Energy Policy* 38/2 (2010): 919-931.

Weiss, Wolfgang. "Chernobyl Thyroid Cancer: 30 Years of Follow-up Overview," *Radiation Protection Dosimetry* 182/1 (2018): 58-61.

Winzer, Christian. "Conceptualizing Energy Security." *Energy Policy* 46 (2012): 36-48.

Withagen, Cees. "Pollution and Exhaustibility of Fossil Fuels." *Resource and Energy Economics* 16/3 (1994): 235-242.

Woo Go, Alchris et al. "Potentials of Agricultural and Agro-industrial Crop Residues for the Displacement of Fossil Fuels: A Philippine Context." *Energy Strategy Reviews* 23 (2019): 100-113.

Xu, Ge et al. "Environmental Risk Perception and Its Influence on Well-being." *Chinese Management Studies* 11/1 (2017): 35-50.

Yan, Qiang et al. "Nuclear Power Development in China and Uranium Demand Forecast: Based on Analysis of Global Current Situation." *Progress in Nuclear Energy* 53/6 (2011): 742-747.

Yang, Rebekah, Ozer, Hasan and L. Al-Qadi, Imad. "Regional Upstream Life-cycle Impacts of Petroleum Products in the United States." *Journal of Cleaner Production* 139 (2016): 1138-1149.

Yankovska, Ganna and Hannam, Kevin. "Dark and Toxic Tourism in the Chernobyl Exclusion Zone." *Current Issues in Tourism* 17/10 (2014): 929-939.

Yardımcı, Okan. "Türkiye Doğal Gaz Piyasası: Geçmiş 25 Yıl, Gelecek 25 Yıl." *Ekonomi Bilimleri Dergisi* 3/2 (2011): 157-166.

Yergin, Daniel. "Ensuring Energy Security." *Foreign Affairs* 85/2 (2006): 69-82.

Yim, Man-Sung and A. Vaganov, Petr. "Effects of Education on Nuclear Risk Perception and Attitude: Theory." *Progress in Nuclear Energy* 42/2 (2003): 221-235.

Yogi Goswami, D. and Kreith, Frank eds. *Handbook of Energy Efficiency and Renewable Energy*. Boca Raton: Crc Press, 2007.

York, Richard and Elizabeth Bell, Shannon. "Energy Transitions or Additions?: Why a Transition from Fossil Fuels Requires more than the Growth of Renewable Energy." *Energy Research & Social Science* 51 (2019): 40-43.

York, Richard. "Do Alternative Energy Sources Displace Fossil Fuels?." Nature Climate Change 2/6 (2012): 441-443.

Yurtseven, İbrahim. "Nehir Tipi Hidroelektrik Santrallerinin Havzalar Üzerindeki Ekohidrolojik Etkileri." *Journal of the Faculty of Forestry Istanbul University* 61/1 (2011): 55-62.

Zeppini, Paolo and CJM Van Den Bergh, Jeroen. "Global Competition Dynamics of Fossil Fuels and Renewable Energy under Climate Policies and Peak Oil: A Behavioural Model." *Energy Policy* 136 (2020): 1-24.

Zhang, Hui et al. "Radiation-driven Migration: The Case of Minamisoma City, Fukushima, Japan, after the Fukushima Nuclear Accident." *International Journal of Environmental Research and Public Health* 11/9 (2014): 9286-9305.

Zhang, Mengyuan et al. "Hemolysis of PM_{10} on RBCs in Vitro: An Indoor Air Study in a Coal-burning Lung Cancer Epidemic Area." *Geoscience Frontiers* 13/1 (2022): 101176.

Zhang, Zuoyi et al. "Design Aspects of the Chinese Modular High-temperature Gas-cooled Reactor HTR-PM." *Nuclear Engineering and Design* 236/5-6 (2006): 485-490.

Zhengfu, Bian et al. "Environmental Issues from Coal Mining and Their Solutions." *Mining Science and Technology (China)* 20/2 (2010): 215-223.

Zhou, Chaoran et al. "Economic Analysis of Two Nuclear Fuel Cycle Options." *Annals of Nuclear Energy* 71 (2014): 230-236.

Zhou, Ying et al. "Impact of Rural Residential Coal Combustion on Air Pollution in Shandong, China." *Chemosphere* 260 (2020): 127517.

Zhou, Yun et al. "Is China Ready for Its Nuclear Expansion?." *Energy Policy* 39/2 (2011): 771-781.

Zmeureanu, Radu and Renaud, Guillaume. "Estimation of Potential Impact of Climate Change on the Heating Energy Use of Existing Houses." *Energy Policy*, 36/1 (2008): 303-310.

"9 Out of 10 People Worldwide Breathe Polluted Air." Last updated date: 25 January 2022, https://www.who.int/news/item/02-05-2018-9-out-of-10-people-worldwide-breathe-polluted-air-but-more-countries-are-taking-action#:~:text=Air%20pollution%20levels%20remain%20dangerously, outdoor)%20and%20household%20air%20pollution.

"Akkuyu NGS İnşaat Projesi," last updated date: 4 April 2022, http://www.akkuyu.com/

"Are There Different Types of Nuclear Reactors?" last updated date: 12 March 2022, https://world-nuclear.org/nuclear-essentials/are-there-different-types-of-reactors.aspx

"Basic Studies in the Field of High-temperature Engineering," last updated date: 14 March 2022, https://www.oecd-nea.org/upload/docs/application/pdf/2019-12/5309-basic-studies.pdf, 96.

"Comparison of Lifecycle Greenhouse Gas Emissions of Various Electricity Generation Sources." Last updated date: 22 January 2022, http://www.world-nuclear.org/uploadedFiles/org/WNA/Publications/Working_Group_Reports/comparison_of_lifecycle.pdf

"Construction Cost Data for Electric Generators Installed in 2019," last updated date: 19 March 2022, https://www.eia.gov/electricity/generatorcosts/

"Electricity Production by Source," last updated date: 9 April, 2022, World, https://ourworldindata.org/grapher/electricity-prod-source-stacked

"Fossil Fuels," last updated date: 30 March, 2022, https://ourworldindata.org/fossil-fuels

"Fuel Comparison." Last updated date: 25 January 2022, https://www.euronuclear.org/glossary/fuel-comparison/

"Geothermal Energy," last updated date: 15 November, 2021, https://www.irena.org/geothermal

"Güncel Türkçe Sözlük," last updated date: 15 January, 2022, https://sozluk.gov.tr/

"International Nuclear and Radiological Event Scale (INES)," last updated date: 5 February 2022, https://www.iaea.org/sites/default/files/ines.pdf

"Meaning of Energy in English," last updated date: 15 January, 2022, https://dictionary.cambridge.org/dictionary/english/energy

"Meaning of Nucleus in English," last updated date: 9 February, 2022, https://dictionary.cambridge.org/dictionary/english/nucleus

"New WHO Global Air Quality Guidelines Aim to Save Millions of Lives from Air Pollution." Last updated date: 26 January 2022, https://www.who.int/news/item/22-09-2021-new-who-global-air-quality-guidelines-aim-to-save-millions-of-lives-from-air-pollution

"Nuclear Energy," last updated date: 5 April, 2022, https://ourworldindata.org/nuclear-energy

"Nuclear Power in the World Today," last updated date: 3 April 2022, https://world-nuclear.org/information-library/current-and-future-generation/nuclear-power-in-the-world-today.aspx

"Nuclear Power in the World Today." Last updated date: 17 January 2022, https://world-nuclear.org/information-library/current-and-future-generation/nuclear-power-in-the-world-today.aspx

"Nuclear Power Plant Construction Costs," last updated date: 20 March 2022, https://www.synapse-energy.com/sites/default/files/SynapsePaper.2008-07.0.Nuclear-Plant-Construction-Costs.A0022_0.pdf

"Oil." Last updated date: 2 February 2022, https://www.bp.com/en/global/corporate/energy-economics/statistical-review-of-world-energy/oil.html#oil-reserves

"OPEC Annual Statistical Bulletin," last updated date: 19 October, 2021, file:///C:/Users/emrah/Downloads/ASB_2020.pdf, s.22.

"Operational & Long-Term Shutdown Reactors" last updated date: 18 March 2022, https://pris.iaea.org/PRIS/WorldStatistics/OperationalReactorsByCountry.aspx

"Projected Costs of Generating Electricity." Last updated date: 18 January 2022, https://iea.blob.core.windows.net/assets/ae17da3d-e8a5-4163-a3ec-2e6fb0b5677d/Projected-Costs-of-Generating-Electricity-2020.pdf, 36.

"Quick Facts on Nuclear Industry Jobs." Last updated date: 30 January 2022, https://www.nei.org/advantages/jobs#:~:text=A%20single%20nuclear%20power%20plant,type%20of%20energy%20generation%20facility.&text=The%20U.S.%20nuclear%20energy%20sector,when%20you%20include%20secondary%20jobs.

"Renewable Energy," last updated date: 7 April, 2022, https://ourworldindata.org/renewable-energy

"Statistical Review of World Energy 2021," last updated date: 19 October, 2021 https://www.bp.com/content/dam/bp/business-sites/en/global/corporate/pdfs/energy-economics/statistical-review/bp-stats-review-2021-coal.pdf, 47.

"Statistics report Key World Energy Statistics 2020." Last updated date: 3 February 2022, https://www.iea.org/reports/key-world-energy-statistics-2020, 13.

"Status of Advanced Light Water Reactor Designs 2004," last updated date: 10 March 2022, https://www-pub.iaea.org/MTCD/Publications/PDF/te_1391_web.pdf

"The 2021 Global Multidimensional Poverty Index (MPI)." Last updated date: 28 January 2022, http://hdr.undp.org/sites/default/files/2021_mpi_report_en.pdf

"Transport," last updated date: 17 February, 2022 https://www.bp.com/en/global/corporate/energy-economics/energy-outlook/demand-by-sector/transport.html

"Unemployment Rate." Last updated date: 28 January 2022, https://ilostat.ilo.org/data/#

"World Energy Outlook 2021." Last updated date: 19 January 2022, https://iea.blob.core.windows.net/assets/4ed140c1-c3f3-4fd9-acae-789a4e14a23c/WorldEnergyOutlook2021.pdf

"World Energy Outlook," last updated date: 12 October, 2021 https://iea.blob.core.windows.net/assets/a72d8abf-de08-4385-8711-b8a062d6124a/WEO2020.pdf, s.138.

"World Nuclear Association, Supply of Uranium," last updated date: 10 March 2022, https://www.world-nuclear.org/information-library/nuclear-fuel-cycle/uranium-resources/supply-of-uranium.aspx

"World Population Growth," last updated date: 31 March, 2022 https://ourworldindata.org/world-population-growth

"World Uranium Mining Production." Last updated date: 3 Febryary 2022, http://www.world-nuclear.org/information-library/nuclear-fuel-cycle/mining-of-uranium/world-uranium-mining-production.aspx

INDEX

www.ingramcontent.com/pod-product-compliance
Lightning Source LLC
LaVergne TN
LVHW050805220926
848485LV00014B/1452/J
* 9 7 8 1 6 8 0 5 3 6 9 2 8 *